THE
Cydonia Controversy

Revised Edition

Mark J. Carlotto

The Cydonia Controversy
(Revised Edition)

Copyright © 2008 by Mark J. Carlotto. All rights reserved. No part of this book may be reproduced or transmitted in any form or by any means without written permission of the author.

ISBN 978-1-4357-5579-6

Cover photo: Colorized version of a Martian landscape imaged by the Spirit rover in 2005. Original black-and-white photo courtesy NASA/JPL/Caltech/Cornell.

Preface

In 1976, a Viking orbiter spacecraft photographed a strange face-like apparition on the surface of Mars. Thirty years later, the origin of the mile-long 'Face on Mars' and a number of other unusual objects in the Cydonia region remain a mystery. Are they artificial — the eroded remains of an extraterrestrial civilization on Mars, or natural — a strange assortment of geological oddities?

My previous book *The Martian Enigmas*, which examined the Viking discoveries, was published while the Mars Global Surveyor was on its way to the Red Planet in the summer of 1997. Soon after the spacecraft arrived, in response to a great deal of public interest in these objects (and some lingering scientific curiosity within NASA as well), a picture was taken of the Face and released to the public within hours. That night an 'enhancement' of the image produced by the Jet Propulsion Laboratory was shown on the evening news. Having downloaded and analyzed the raw data earlier in the day, the image I saw on TV did not look anything like the one in front of me on my laptop computer. More of a corruption of the data than an enhancement, the image, released without comment by JPL/NASA and accepted without question by the media, to the delight of critics, 'settled' the debate over the Face once and for all.

Since then several books have been published, claiming NASA has proof of an extraterrestrial intelligence existing on Mars, that it is a part of a larger 'cosmic conspiracy', and is withholding this and other secrets from the public. My reason for writing *The Cydonia Controversy* is to present a more specific hypothesis: that the Face and other objects are artificial constructions that were built long ago, but appear as they do today, highly eroded and, in some places, covered with sand and debris, as a result of their long-term exposure to the Martian environment. As a result, it is not at all surprising that even after extensive high-resolution imaging of this area by MGS and other spacecraft, no obvious indications of artificiality, no buildings or roads, are evident. Yet, there are many unusual characteristics about these objects and this area on Mars that remain unexplained: symmetries, alignments, and other subtle patterns of organization that seem to tip the balance of explanation toward intelligent design over what has been termed enigmatic geology.

My other reason for writing this book is to place the thirty-year controversy over the Face on Mars within a broader scientific and political context, rather than one of cover-up and conspiracy. Chapter 1 reviews the controversy and raises a number of key questions, which are echoed throughout the book. Chapters 2 and 3 trace the origin and evolution of our interest in extraterrestrial life and other 'worlds' from myth and legend, through the Renaissance, the Martian canal controversy, UFO sightings in the 1950's, and on to the Viking mission. Chapter 4 is about the discovery of the Face by NASA, its rediscovery by two engineers working at NASA a few years later, and the investigation that followed. Evidence from the Viking mission in support of artificiality is presented in Chapter 5, and its plausibility, given what we know (or think we know) about Mars, is considered in Chapter 6. The politics surrounding the Face and Cydonia, meetings with officials at NASA and JPL, and events leading up to the MGS mission are covered in Chapter 7. Chapter 8 presents a detailed analysis of the April 1998 MGS image of the Face. Chapter 9 analyzes later imagery from MGS and the Mars Odyssey spacecraft that show definite indications of symmetry and architectural design in a number of objects, including the Face. Chapter 10 summarizes the scientific evidence supporting the case for artificiality in Cydonia, and reviews it in the context of the extraterrestrial life debate — one that began thousands of years ago in ancient Greece. Chapter 11 concludes by suggesting several new directions for future investigations.

With all of the challenges facing our planet, a manned mission to Mars may not seem like a priority. But if our race is to flourish and not merely survive, we need new frontiers. Mars is the closest and potentially the most habitable planet in our solar system. When we get there what will we find? A lifeless planet as some planetary geologists believe? A cold dry desert populated by tiny microbes and lichens as new evidence suggests? Or a world that was once visited, perhaps even inhabited, by beings like us?

The Face on Mars may be the key to understanding who we are — of what once was, or might be, or what simply is. Is it an ancient monument, a faint and distant echo from the past, a reminder of an earlier age, long forgotten, or an extraterrestrial 'calling card' beckoning us to come and explore, or just a pile of rocks?

The dreamer in me believes the Face is artificial, but the scientist requires proof. Join me in a journey into the future by way of our past.

Mark J. Carlotto
Gloucester, Massachusetts
September 2008

Contents

One — Introduction   1

Two — From Myth to Science   5
    Mythological Origins   5
    Sumerians and Babylonians   7
    The Greeks   10
    The Renaissance   14
    Early Speculation about Life on the Moon and Planets   16
    Toward a Scientific Theory of Extraterrestrial Life   19
    The Canals on Mars   21
    Sources   23

Three — Terrestrial Intelligence   25
    Unidentified Flying Objects   25
    The Search for Extraterrestrial Intelligence   28
    Alternative SETI Concepts   30
    Photoreconnaissance of the Moon   31
    The Brookings Report   36
    Scientific Study of UFOs   37
    UFOs in Space   38
    The Search for Life on Mars Resumes   41
    Sources   41

Four — Contact   43
    Terrestrial Mind Set   43
    Rediscovery   45
    A City on Mars?   47
    Solstice Hypothesis   49
    The Independent Mars Investigation   49
    Evidence of Artificiality   51
    Research Methodology   52
    Preliminary Conclusions   53
    Planetary Scientists Respond   55
    Sources   56

Five — The Evidence   57
    A Closer Look   58
    A Trick of Light and Shadow?   59
    Measuring Artificiality   62
    Strange Geomorphology   65
    Tetrahedral Geometry   66
    Testing For Random Geology   68
    Reviewing the Evidence   70
    Extraordinary Claims Require Extraordinary Evidence   77
    Sources   79

Six — City by the Sea	**81**
The Crustal Dichotomy	81
A Paleo-Ocean on Mars?	83
Mars Past, Present, and Future	86
Evidence of Life on Mars	91
A Theory of Life	93
Sources	95
Seven — The Politics of Cydonia	**97**
Mars Observer	97
The McDaniel Investigation	101
How Alien is Alien?	104
The Ethics of Scientific Debate	105
The Mars Orbiter Camera	107
Mars Global Surveyor	108
Targeted Observations	110
Sources	112
Eight — New Evidence of Artificiality in Cydonia	**113**
The Face Disappears	115
Restoring the Face	116
Comparing Facial Features	118
Correcting Geometrical Distortions	120
Announcing Our Findings	121
Other Anomalies in Cydonia	125
NASA's Promise	128
Sources	132
Nine — On the Threshold of a Dream	**133**
Evidence of Water and Primitive Life on Mars?	134
On The Symmetry of the Face	136
A Legal Ultimatum	139
Analysis of the April 2001 Image of the Face	140
More on Symmetry	148
Sources	152
Ten — Toward a Synthesis of Science and Myth	**153**
The Cydonia Controversy and the Extraterrestrial Life Debate	153
Extraterrestrial Archaeology	158
Terrestrial Connections	167
Return to Myth	172
Sources	174
Eleven — Epilog	**177**
A Signature of Artificiality	177
Patterns of a Technological Intelligence	179
Reconsidering Basic Assumptions	185
Sources	189
Index	**191**

One — Introduction

> When a distinguished elder scientist states something is impossible, he is almost always wrong. — Arthur C. Clarke

Late in the summer of 1975, two Viking spacecraft, each consisting of an orbiter and a lander, were launched toward the planet Mars. The goal of the mission was to search for life, microbial life thought to exist in the red Martian soil. Eleven months later, the first spacecraft reached the Red Planet, achieved orbit, and began its search for a landing site. On July 20, 1976 the lander separated, descended, and successfully touched down on the surface. As it began sending television pictures and other scientific data back to NASA's Jet Propulsion Laboratory in Pasadena, California, the orbiting mother ship continued its mapping mission, preparing for the arrival of the second Viking spacecraft, due a little over a week later.

Early in the morning of July 25, while searching for a second landing site, the orbiter photographed an enormous face-like rock formation in the Cydonia region of Mars. After no more than a cursory examination of the picture, JPL dismissed the 'Face' as an optical illusion — as a geological feature that happened to look like a face because of the way it was lit by the sun. Viking project scientist Gerry Soffen informed the press that the Face disappeared in an image taken later in the day. In the following weeks as attention focused on the two landers and their search for microbes on the surface, the Face on Mars was forgotten.

A few years later, Vincent DiPietro and Gregory Molenaar, engineers at NASA's Goddard Space Flight Center, just outside of Washington DC, came across the same picture of the Face in the Viking photo archives and decided to learn more about it. Soffen had said that another picture taken later in the day showed nothing unusual. However, DiPietro and Molenaar could not find that picture because, as they soon discovered, later on July 20, Viking was thousands of miles away from Cydonia. They continued to search and finally did find another image taken 35 days later. To their amazement, the Face was still there.

DiPietro and Molenaar's finding was totally ignored by JPL and the planetary science community. Gradually over time, however, others began to take notice. One of the first was science writer Richard Hoagland, who realized that the Face was not an isolated formation, but happened to be near a group of unusual-looking geometrical features that reminded him of Paolo Soleri's archologies — an idea for housing large urban populations in three-dimensional pyramidal structures. With the help of anthropologist Randy Pozos, Hoagland organized a group known as the Independent Mars Investigation Team to examine the Face and the other features in greater detail. Over the next few months, the group, which included DiPietro and Molenaar, physicists Lambert Dolphin and John Brandenburg, artist James Channon, and others, found additional evidence that Viking had discovered archaeological ruins on Mars.

The response of the planetary science community was surprising. When asked about the Face, geologists thought other landforms on Mars were more interesting. Gerry Soffen, the Viking project scientist who informed the press the Face disappeared in a later image, was simply not interested. The late astronomer Carl Sagan, then a leader in the search for extraterrestrial intelligence, was of the opinion that any serious study of the Face was a waste of time.

Why would JPL and the planetary science community be unwilling to consider the possibility of artificial structures on Mars? After all, Viking was sent to Mars to look for life. Why did Soffen, a senior project scientist, claim that in a second image taken a few hours later, the Face disappeared? Moreover, how could he make such a statement knowing later in the day Viking was thousands of miles away? Why didn't JPL bother to mention that when a second image was finally taken of the Face thirty-five days later, under different lighting conditions, it was still visible and so could not be an optical illusion? Moreover, how could JPL scientists who were trained photo-interpreters miss the other nearby objects, some even stranger than the Face?

From the beginning, JPL has insisted that the Face and other objects in Cydonia are simply odd-looking rock formations, not unlike those seen in the American southwest. They argue that these structures must have formed by the chance interplay of a variety of natural processes such as volcanism, tectonism, catastrophic flooding, mass wasting, freezing and thawing, wind erosion and deposition, fluvial erosion, glaciation, and meteoric impact. They are sure that suitable conditions did not last long enough for intelligent life to develop on Mars, and that the very idea of a humanoid face on Mars contradicts everything they know (or think they know) about Mars. In effect, they are saying that because the Face cannot be there, it is not. But do planetary scientists know enough about Mars to state definitively that the Face cannot be artificial?

From the middle of the 18th century to the beginning of the 20th century, Mars was believed to be a planet much like Earth. Sir William Herschel discovered that it is tilted on its axis, and so like Earth, has seasons. He observed changes in the polar caps which he thought were caused by the seasonal melting of snow and ice by the sun. In the 19th century as the telescope improved, astronomers began to notice changes in the tones and colors of the Martian surface. Some thought the dark areas were wet soil, others believed they were seas. Giovanni Schiaparelli saw linear features, which he thought, were channels. Percival Lowell later saw them as canals — artificial waterways constructed by an advanced race of Martians to save their dying planet. Around the turn of the 20th century, spectrographic measurements of the Red Planet revealed that it had very little of the two prerequisites for life: water and air. The canals were shown to be an optical illusion caused by the tendency of our eyes to see poorly resolved patterns as linear features.

Early unmanned space probes in the mid 1960's initially confirmed Mars was a much-less hospitable place than originally thought. Mariners 4, 6, and 7 flew past the Red Planet, each returning a handful of photographs showing a heavily cratered surface that looked more like the Moon than the Earth. But a few years later, our view changed again when Mariner 9 orbited Mars and imaged sights that had never before been seen on another planet: enormous volcanoes, vast canyon systems, and extensive networks of channels and tributaries. Orbital photography showed Mars once had liquid water on its surface, and so must have had a thick atmosphere as well.

In 1971, Viking was sent to Mars to follow-up on these startling revelations. The fact that the planet probably once had water and an atmosphere did not change the reality of its present environment — one that is hostile to life. So, instead of searching for life on the surface, Viking sought out microbes buried in the Martian soil. At both landing sites, three on-board biology experiments returned positive indications of life. But because a fourth experiment did not find a sufficient amount of organic material in the soil, most, but not all, JPL scientists decided the biology experiments were responding to the highly oxidized Martian soil. They concluded there is no life on Mars.

For the next twenty years, planetary scientists believed Mars is a dead planet, and has been for billions of years. When in 1989 a group of British scientists claimed they had found evidence of organic compounds in a 200 million-year-old meteorite thought to be from Mars, the planetary science community ignored them. But in 1996, after NASA announced it had found evidence of microbes in a much older Martian meteorite, one about 4.5 billion years old, scientific opinion reversed itself once again about the possibility of life on the Red Planet.

After more than twenty years, in response to the growing public interest in Cydonia, the Mars Global Surveyor re-photographed the Face on April 5, 1998. But instead of resolving the controversy, this new photo seemed to raise even more questions. Within hours, JPL posted a 'contrast-enhanced' image on its web site that was processed in such a way as to make the Face appear flat and featureless. Unaware of JPL's processing of the photo, most viewers came to the conclusion that night on seeing the picture in the evening news that the Face on Mars was what JPL had claimed it was all along — just a pile of rocks. On the same day the picture was taken, the media pronounced the case for the Face closed.

In the following months, after downloading the original unenhanced imagery from the JPL web site, and carefully restoring, enhancing, and correcting for various distortions, a completely different impression of the Face emerged. Although the new MGS image did show the Face to be heavily eroded, there was still much about it that was unusual. Unlike nearby mesas, features clearly of natural origin, the platform surrounding the Face seemed highly symmetrical. What appeared as two faint lines on the top of the head in the lower resolution Viking images were resolved by MGS into a pair of features that looked as if they were cut into the side of the Face, much like access ramps, which would allow someone at ground level to reach the top of the Face platform. The biggest surprise was finding 'nostrils' and lip-like structures along the lateral centerline of the Face — just the kind of facial details one would expect to find on a representation of a humanoid head. Remarkably, the Face seemed to pass the Lowellian litmus test. Unlike Lowell's canals that disappeared in better resolution imagery, the Face turned out to be even more interesting up close.

When these findings were presented at the *1998 Spring Meeting of the American Geophysical Union* in Boston, JPL's chief MGS scientist, Arden Albee, became upset. It was clear that JPL's enhanced image was a gross distortion of the Face, but was it an honest mistake, done in haste to make the picture available to the public and the media, or was it an attempt by certain individuals to put the matter to rest once and for all?

In April 2001, MGS captured a fully illuminated, high-resolution, overhead image of the Face. Seen in its entirety for the first time, the formation shows clear evidence of bilateral symmetry and rectilinear geometry — strong indicators of artificiality. Yet, JPL and the planetary community continue to insist the Face is natural, belittling arguments to the contrary.

One has to wonder, if JPL is so sure about the Face, why have they repeatedly misinformed and deceived the public? Why do they say there is no credible evidence the Face is artificial, disregarding scientific analysis that indicates otherwise? Why do they ignore findings published in peer-reviewed scientific and technical journals that show the Face is not an optical illusion, that it contains fine-scale detail that cannot be explained, and is but one of a handful of objects, all in the same general area, that also appear to be highly unusual in their own right? Why have they distorted Viking and MGS imagery to make the Face look like an ordinary geological formation. And why has JPL, the organization responsible for exploring Mars, sidestepped the Cydonia controversy by letting the media and the 'court of public opinion' decide the fate of the Face? Why are they using the public, and a misinformed public at that, to decide a matter as important as the potential discovery of archaeological ruins on Mars?

Two — From Myth to Science

> To be ignorant of what occurred before you were born is to remain always a child. For what is the worth of human life, unless it is woven into the life of our ancestors by the records of history? — Cicero

The possibility of artificial structures on Mars is part of the larger question of whether or not there is other intelligent life in the universe. Despite the current interest in SETI — the search for extraterrestrial intelligence using radio telescopes popularized by Carl Sagan in his book *Contact* — the idea of extraterrestrial life is not new, as some may think. Rather, it is an age-old question that has been asked time and time again. It was debated late in the 19th century in the context of the controversy over the Martian canals. Before that was the question of whether there was life on the Moon, and the other planets in our solar system. During the late Middle Ages and Renaissance, Christian scholars considered the possibility of other worlds — the 'plurality of worlds', which was first introduced by the earliest of the Greek philosophers.

Although Western scientific and philosophical thought regarding extraterrestrial life can be traced back to ancient Greece, the origin of the question itself is much older — one deeply rooted in myth and legend. In this chapter, we outline the evolution of the idea of extraterrestrial intelligence from its mythological origins through early scientific thought at the beginning of the 20th century in order to provide a historical context for evaluating the Cydonia discoveries, and a framework for considering the implications of the discovery of artifacts on Mars later in the book.

Mythological Origins

The earliest references to other beings and other worlds are found in the Old Testament and ancient Sumerian texts. The *Book of Genesis* tells us long ago [1]

> There were giants in the Earth in those days; and also after that, when the sons of God came in unto the daughters of men, and they bare children to them, the same became mighty men which were of old, men of renown.

According to the noted biblical scholar Zecharia Sitchin, the term 'giants' is one translation of the original Semitic term 'Nefilim', which literally means 'those who were cast down upon Earth'. Sitchin's alternative translation of this verse from Genesis is [2]:

> The Nefilim were upon the Earth, in those days and thereafter too, when the sons of the gods cohabitated with the daughters of Adam, and they bore children unto them. They were the mighty ones of Eternity — The people of the Shem.

'Shem', which is commonly translated as 'name', is also the term for a commemorative monument, or stela representing 'stones that rise' and 'objects that give off light'. This suggested to Sitchin that 'shem' might be a conveyance of some kind, like a rocket, and that the Nefilim were beings that had come to Earth from the heavens.

Certain events in the Bible seem to corroborate this 'extraterrestrial' interpretation of the gods. The *Book of Ezekiel* states:

> Now as I beheld the living creatures, behold one wheel upon the Earth by the living creatures, with his four faces. The appearance of the wheels and their work was like unto the colour of a beryl: and they four had one likeness: and their appearance and their work was as it were a wheel in the middle of a wheel. When they went, they went upon their four sides: and they turned not when they went. As for their rings, they were so high that they were dreadful; and their rings were full of eyes round about them four. And when the living creatures went, the wheels went by them: and when the living creatures were lifted up from the Earth, the wheels were lifted up.

And from the *Book of Kings*:

> And it came to pass, as they still went on, and talked, that, behold, there appeared a chariot of fire, and horses of fire, and parted them both asunder; and Elijah went up by a whirlwind into heaven.

That these biblical excerpts seem not unlike those of modern day reports of unidentified flying objects led Erich von Daniken in his 1968 book, *Chariots of the Gods?*, to propose the idea that the gods were 'ancient astronauts'. After all, von Daniken reasoned, "If our own space travelers happen to meet primitive peoples on a planet one day, they too will presumably seem like 'sons of heaven' or 'gods' to them" [3]. Because of the rather sensationalistic way in which the material was presented, von Daniken's ideas were brutally attacked by the scientific establishment. Yet, the idea was not really that outrageous.

A few years earlier in a paper entitled "Direct contact among galactic civilizations by relativistic interstellar flight," Carl Sagan had considered the very same possibility. Concerned with the difficulty in interpreting myth and legend, he stated that under certain circumstances an encounter with an alien civilization might be recorded in a reconstructable manner if "the account is committed to written record soon after the event, a major change is effected in the contacted society by the encounter, and no attempt is made by the contacting civilization to disguise its exogenous nature" [4]. He goes on to say that there are legends that satisfy these criteria, which deserve further study, one being "the Babylonian account of the origin of the Sumerian civilization by the Apkallu, representatives of an advanced, nonhuman and possibly extraterrestrial society."

In his book, *The 12th Planet*, Sitchin argues that the Nefilim were none other than the 'olden gods' of the Sumerians, the 'gods of Earth'. The head of this, the original pantheon of gods was Anu, who resided in 'heaven'. He had two sons, Enlil and Ea. Second to Anu, Enlil, the 'lord of the airspace' ruled Earth from the ancient city of Nippur, located in present day Iraq. Below Enlil in rank, was Ea (also known as Enki), who played a seminal role in human affairs — for it was Enki together with his half-sister, the goddess Ninhursag (also called Ninti) who, according to Sumerian legends, created Mankind, and Ea who later warned Utnapishtim, the Biblical Noah, about the impending Flood, as told in the *Epic of Gilgamesh*. Other gods included Enlil's son, Nannar (Sin), who ruled over the Sumerian city state of Ur, his daughter, Inanna (Ishtar) — the goddess of love (known as Aphrodite to the Greeks and Venus to the Romans), her sister, Ereshkigal, who ruled over the 'lower world', and Ereshkigal's consort, Nergal, the name the Babylonians later used to refer to the planet Mars. In all, there were twelve principal gods on Earth, and 900 lower gods known as 'Annunaki' of which 600 were on Earth and 300 remained in 'heaven'.

If the Nefilim were extraterrestrials, where did they come from? Mesopotamian cylinder seals suggest one possibility. According to the Sumerian creation epic *Enuma Elish*, an object from outside our solar system (Marduk) collided long ago with a planet known as Tiamat located between Mars and Jupiter. The largest fragment became the Earth, with the remnants left to form the asteroid belt. Marduk, also known as Nibiru, ended up in a highly elliptical retrograde orbit extending from well outside that of Pluto to the asteroid belt. Drawing on Mesopotamian and biblical sources, Sitchin believes Nibiru to have an orbital period of 3600 years. This interpretation of the Sumerian creation myth is consistent with depictions in several cylinder seals of a star-like object (thought to be the Sun) surrounded by eleven other objects — corresponding to the known nine planets, the Moon, and Nibiru. It is Sitchin's contention that this tenth, yet to be discovered planet, Nibiru, is the home of the Nefilim. By analyzing a variety of ancient sources, he estimates the Nefilim came to Earth roughly 445,000 years ago.

Sumerians and Babylonians

The Sumerian civilization was a highly developed and, in many ways, a truly modern civilization, which suddenly appeared in the Near East around 3800 B.C. From the study of thousands of clay tablets, scholars have come to the conclusion that the Sumerian culture was, in essence, not much different from our own. They had an extensive practical knowledge of medicine (including surgical procedures), agriculture (e.g., shade tree gardening, irrigation), pottery and metallurgy (making use of the region's rich petroleum resources to fire their kilns), mathematics, and astronomy. They used a base-60 number system, which allowed them to count easily into the millions, and to work with fractions — a number system that was far superior to later ones (e.g., Roman numerals). They divided the circle into 360 degrees, and the year into twelve months. They named hundreds of stars, and originated the twelve signs of the zodiac. The Sumerians knew the Earth was spherical, not flat, as later thought by the Egyptians and the Greeks. Yet, according to Sitchin, "the perplexing fact about this is that to this day the scholars have no inkling who the Sumerians were, where they came from, and how and why their civilization appeared."

It has been shown that cuneiform — the style of wedge-shaped writing on clay tablets used by the Babylonians — was derived from an earlier pictographic script, like Egyptian hieroglyphics, that was used by the Sumerians. But, in an ancient text found in the archaeological remains of the city of Nineveh (near modern-day Mosul), this statement made in the second millennium B.C. by the Assyrian king Ashurbanipal implies that the Sumerians did not invent it:

> The god of scribes has bestowed on me the gift of the knowledge of his art. I have been initiated into the secrets of writing. I can even read the intricate tablets in Shumerian; I understand the enigmatic words in the stone carvings from the days before the Flood[1].

Paleoarchaeological evidence shows that man became a farmer in the Near East around 11,000 B.C., and shortly thereafter began to domesticate plants and animals. Pottery appeared around 7500 B.C., followed by metallurgy a millennium and a half later. That man had developed a written language before the mythical Flood, before the invention of agriculture when we were still hunter-gathers, a written language sophisticated enough be used thousands of years later by the Babylonians, would seem unlikely. Yet, etymological evidence strongly suggests that a written language did exist as far back as 15,000 B.C.

Figure 1 A depiction of a Sumerian cylinder seal showing a winged object between Earth and Mars. The Earth and Moon are represented by seven circles to the left, and the crescent to the right of the god on the left. Mars is symbolized by the six-pointed star to the left of the god on the right.

---

[1] Translated from one of over 25,000 clay tablets excavated at the site of the Library of Nineveh [2].

Equally mysterious is the origin of their knowledge of astronomy. The Babylonians believed in the astrological significance of celestial events, for example,

> If Mars comes close to the Great Twins [the constellation Gemini]: the king will die and there will be hostilities.

The *Enuma Anu Enlil* and other ancient texts contain thousands of such omens [5]. It was the chief duty of Babylonians astronomers to develop tables of celestial events (emphemerides) such as heliacal risings (the day when a planet, star or constellation is first seen at sunrise or sunset), conjunctions (when two celestial bodies have the same longitude on the celestial sphere), occultations (when one celestial body such as the Moon occludes another), and many others. According to the Greek historian Herodotus, Thales, who is considered to be the father of Greek science and philosophy, used such information to end a war by predicting a solar eclipse in 585 B.C.

For centuries, Babylonian astronomers recorded celestial observations in texts known as the *Astronomical Diaries* [5]. Although it would seem these observations should suffice for prognostication purposes (except perhaps when it was cloudy), they also had in their possession 'procedure texts' — ancient texts containing algorithms for predicting the date of celestial events. N.M. Swerdlow, an astronomer at the University of Chicago, has argued that Babylonian astronomers could have derived the underlying numerical models contained in procedure texts from data in the *Astronomical Diaries*. The problem with this explanation is that the accuracy of the algorithms in the procedure texts is better than the data from which they were supposedly derived! Some scholars believe the *Astronomical Diaries* were administrative records and were not used for astronomical purposes. Others think that Babylonian astronomers did use the procedure texts to compute emphemerides, but were ignorant of the theories on which they were based. If this is so, where did this knowledge come from?

The Sumerians, and later the Babylonians, possessed a great deal of practical information about agriculture, medicine, astronomy, and other fields, but lacked anything resembling the scientific knowledge that we have come to know. Although the Babylonians could predict the time of a particular celestial event, they could not predict the continuous motion of a celestial body, as the Greeks would later attempt to do because they did not have a physical model of the phenomenon. Their knowledge was a practical one of cause and effect. That the Eastern peoples knew so much without having a deeper scientific understanding of the underlying principles is hard to reconcile, unless one is willing to consider the possibility of an earlier and/or outside influence.

In considering similarities in architecture, language, myth, and other characteristics between otherwise disparate cultures divided by time and geography, scholars have come to conclusion that the earliest cultures in recorded history were seemingly influenced by an even earlier prehistoric culture. Writers like Charles Berlitz and Graham Hancock believe these early cultures inherited their knowledge from an advanced civilization that was wiped out by the Flood — the mythical 'Atlantis'. Von Daniken and Sitchin have gone a step further, proposing that the original source of their knowledge was extraterrestrial in nature. Sitchin's account of the origin of the Sumerian and other early civilizations, and similar theories are particularly appealing as they also provide a possible explanation for why we are, and seemingly always have been, curious about extraterrestrials.

The Greeks

Interestingly enough, the origin of Western thought concerning the existence of other worlds, and the possibility of life on these worlds, coincided with the birth of Greece philosophy and science in the 6th century B.C. It occurred at a time when early Greek scholars were traveling to places like Egypt and Babylon to study mathematics and astronomy. According to John Burnet, an expert in early Greek philosophy, "It cannot be an accident that philosophy originated just at the time when communication with these two countries was easiest, and that the very man [Thales] who was said to have introduced geometry from Egypt is also regarded as the first philosopher" [6].

The ancient Egyptians believed the world was a circular disk floating on a great ocean inside a hemisphere of shining stars. This was the point of departure for Greek cosmological thinking. The philosopher Anaximander transformed the flat Earth of the Egyptians into a cylinder. Pythagoras (who is associated with the famous theorem that the sum of the squared lengths of the sides of a right triangle is equal to the squared length of the hypotenuse) later made it spherical. Initially, Earth was at the center of the Greek cosmos. The Pythagorean School of Greek philosophy had it revolve around a central fire (not the sun). Later, Aristarchus put the sun in the center, with Earth and the other planets revolving around it, only to have Hipparchus and Ptolemy put it back in the center.

In the process of developing their cosmology, Greek philosophers considered the idea of other worlds. Epicurus, in his *Letter to Pythocles*, defined the 'world' as "a circumscribed portion of the universe, which contains stars and Earth and all other visible things, cut off from the infinite" [7]. Interestingly, the Greeks wondered not whether there were other inhabited planets (other 'Earths') in our world (what today we would call our universe), but rather whether there were other worlds (other universes). Many believed there were an infinite number of worlds, an idea known as 'innumerable worlds'.

Some thought that other worlds had existed in the past. According to Cicero, "Anaximander's opinion was that there were gods who came into being, rising and passing away at long intervals, and that these were the innumerable worlds" [6]. Xenophanes stated, "All human beings are destroyed when the Earth has been carried down into the sea and turned to mud. This change takes place for all the worlds." The idea of past worlds was probably motivated by legends brought to Greece from the East of the great Flood and other calamities as in this famous passage from Plato's *Timaeus*, where Critias recounts a tale told to his grandfather by Solon, a contemporary of Thales [8]:

> On one occasion, wishing to draw them on to speak of antiquity, he [Solon] began to tell about the most ancient things in our part of the world... Thereupon one of the [Egyptian] priests, who was of a very great age, said: O Solon, Solon, you Hellenes are never anything but children, and there is not an old man among you. Solon in return asked him what he meant. I mean to say, he replied, that in mind you are all young; there is no old opinion handed down among you by ancient tradition, or any science, which is hoary with age. And I will tell you why. There have been, and will be again, many destructions of mankind arising out of many causes; the greatest have been brought about by the agencies of fire and water, and other lesser ones by innumerable other causes.

Others believed that multiple worlds coexisted in time. Around the beginning of the 5th century B.C., the idea that the world is composed of tiny indivisible particles known as atoms was proposed by Leucippus and Democritus. According to Theophrastus [6],

> He [Leucippus] assumed innumerable and ever-moving elements, namely, the atoms. And he made their forms infinite in number, since there was no reason why they should be of one kind rather than another, and because he saw that there was unceasing becoming and change in things. He held, further, that 'what is' is no more real than 'what is not', and that both are alike causes of the things that come into being for he laid down that the substance of the atoms was compact and full, and he called them 'what is', while they moved in the void which he called 'what is not', but affirmed to be just as real as 'what is'.

The noted historian, Steven J. Dick attributes Greek interest in other coexistent worlds to be a direct consequence of their belief in atomism. The atomist view of the world was one filled with atoms that were perpetually in motion, colliding, sticking together, and coming apart. Extending this view to the universe it seemed reasonable that worlds were constantly in a state of flux, being created, existing for some period of time, and finally destroyed. Again, according to Theophrastus,

> He [Leucippus] says that the All is infinite, and that it is part full, and part empty. These (the full and the empty), he says, are the elements. From them arise innumerable worlds and are resolved into them.

He then goes on to describe how worlds come into being:

> There were borne along by 'abscission from the infinite' many bodies of all sorts of figures 'into a mighty void', and they being gathered together produce a single vortex. In it, as they came into collision with one another and were whirled round in all manner of ways, those which were alike were separated apart and came to their

likes. But, as they were no longer able to revolve in equilibrium owing to their multitude, those of them that were fine went out to the external void, as if passed through a sieve; the rest stayed together, and becoming entangled with one another, ran down together, and made a first spherical structure.

In the middle of the 4th century B.C., Epicurus summarized the case for innumerable worlds in this way [6]:

> There are infinite worlds both like and unlike this world of ours. For the atoms being infinite in number, as was already proved, are borne on far out into space. For those atoms which are of such nature that a world could be created by them or made by them, have not been used up either on one world or a limited number of worlds... So that there nowhere exists an obstacle to the infinite number of worlds.

Epicurus believed that because the world was a part of the infinite there must be an infinite number of other worlds. Plato and Aristotle thought differently. From Plato's *Timaeus* written in 360 B.C. [8]:

> Are we right in saying that there is one world, or that they are many and infinite? There must be one only, if the created copy is to accord with the original. For that which includes all other intelligible creatures cannot have a second or companion; in that case there would be need of another living being which would include both, and of which they would be parts, and the likeness would be more truly said to resemble not them, but that other which included them. In order then that the world might be solitary, like the perfect animal, the creator made not two worlds or an infinite number of them; but there is and ever will be one only-begotten and created heaven.

Where Plato's belief in a single world was based on his equating uniqueness with perfection, Aristotle's was based in large part on the concept of 'natural place'. To Aristotle, the world consisted of four elements: earth, air, fire, and water. Each element sought its natural place in the world. Earth moved naturally to the center, fire to the periphery, and air and water in between. Aristotle assumed, for the sake of argument, if there was more than one world, "these worlds, being similar in nature to ours, must all be composed of the same bodies as it" [9]. He then reasoned that

> The particles of earth, then, in another world move naturally also to our centre and its fire to our circumference. This, however, is impossible, since, if it were true, Earth must, in its own world, move upwards, and fire to the centre; in the same way the Earth of our world must move naturally away from the centre when it moves towards the centre of another universe. This follows from the supposed juxtaposition of the worlds. For either we must refuse to admit the identical nature of the simple bodies in the various universes, or, admitting this, we must make the centre and the extremity one as suggested. This being so, it follows that there cannot be more worlds than one.

With Earth at the center of the world, fire at the circumference, and air and water in between according to Aristotle,

> ...it is evident not only that there is not, but also that there could never come to be, any bodily mass whatever outside the circumference. The world as a whole, therefore, includes all its appropriate matter, which is, as we saw, natural perceptible body. So that neither are there now, nor have there ever been, nor can there ever be formed more heavens than one, but this heaven of ours is one and unique and complete.

As a consequence of natural place, Aristotle argued there can be nothing beyond the world, not even void:

> It is therefore evident that there is also no place or void or time outside the heaven. For in every place body can be present; and void is said to be that in which the presence of body, though not actual, is possible; and time is the number of movement. But in the absence of natural body there is no movement, and outside the heaven, as we have shown, body neither exists nor can come to exist. It is clear then that there is neither place, nor void, nor time, outside the heaven.

Following the Golden Age of Greek philosophy, through the rise and fall of the Roman Empire, the Middle Ages, and into the Renaissance, Aristotle's belief in one world would remain the focal point in the debate over the possibility of other worlds.

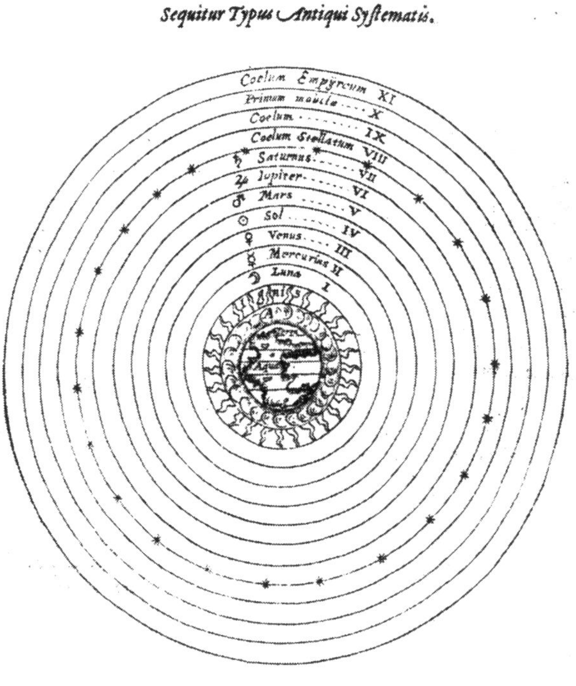

Figure 2 Aristotle's Cosmos from Boulliau's *Philolaus* (Amsterdam 1639). In Aristotle's model [11], the sun revolves around the Earth. The orbits of the Moon, Mercury, and Venus lie between the Earth and sun while those of Mars, Jupiter, and Saturn are beyond. Beyond the planets lie the stars and sky.

The Renaissance

After the fall of the Roman Empire, Muslim scholars preserved and expanded upon Greek science and philosophy until it was reintroduced into Europe in the 12th century. Soon after, the plurality of worlds debate resumed. Albertus Magnus believed the question of other worlds was one worthy of study [9]:

> Since one of the most wondrous and noble questions in Nature is whether there is one world or many, a question that the human mind desires to understand *per se*, it seems desirable for us to inquire about it.

Early Christian scholars such as Roger Bacon mostly reiterated Aristotle's ideas about natural place. However, there was a problem: In contrast to the Christian belief in the end of the world, Aristotle's world was eternal. The Dominican friar Thomas Aquinas attempted to reconcile Aristotle's universe with Christian teachings. But as scholars began to question these teachings, the idea of multiple worlds was ultimately condemned along with hundreds of other widely held beliefs at the time by the Catholic Church in 1277.

As time passed, Christian scholars began to modify Aristotle's model of the universe, and in doing so, gradually removed logical and religious barriers to the possibility of other worlds. Early in the 14th century, William of Ockham argued that God could create worlds with different natures, and as a result could create more than one world. Where Aristotle had defined the concept of natural place in an absolute sense, Ockham redefined it so as to be relative to a given world. Unfortunately, opposition to Ockham's ideas was so great that he was expelled from the Franciscan order and excommunicated by the Church.

A half-century later, Nicole Oresme went a step further arguing that two worlds far enough removed need not have any relation at all to one another, opening the door to the possibility of more than one world. Where Aristotle had said that "there is neither place, nor void, nor time, outside the heaven" Oresme believed God "is infinite in His immensity, and if several worlds existed, no one of them would be outside Him nor outside His power" [9]. Although he argued that other worlds were possible in theory, in deference to Church authority, Oresme, like other scholars of the time, would conclude their argument with a statement to the effect that "there has never been nor will there ever be more than one world," just to be safe.

Gradually over time, Aristotle's concept of natural place was turned inside-out. Early in the 15th century, Nicholas of Cusa advanced a completely different view of the universe, one in which its "center is everywhere and its circumference nowhere" [9]. According to Steven Dick, this was a turning point as people began to think of the celestial bodies themselves as worlds. It was also at about this time that the idea of extraterrestrial life entered into the discussion.

Since Epicurus the belief in a plurality of worlds was strongly rooted in the 'principle of plenitude' — the idea that if something is possible, it will eventually be realized in nature. In 50 B.C., the Roman poet Lucretius put it this way: "when matter abundant is ready there, when space on hand, nor object nor any cause retards, no marvel 'tis that things are carried on and made complete" [11]. This, in turn, motivated further speculation on life beyond our world,

> I say, again, again, 'tmust be confessed there are such congregations of matter otherwise, like this our world which vasty ether holds in huge embrace.

and even the idea of others like ourselves on those worlds,

> 'Tmust be confessed in other realms there are still other worlds, still other breeds of men, and other generations of the wild.

Probably the first to touch upon the idea of life on other worlds since Lucretius, Cusa believed the Sun and Moon to be other worlds like Earth, and speculated on the nature of the life on these worlds [9]:

> It may be conjectured that in the area of the sun there exist solar beings, bright and enlightened intellectual denizens, and by nature more spiritual than such as may inhabit the Moon — who are possibly lunatics — whilst those on Earth are more gross and material.

He went on to say that:

> Rather than think that so many stars and parts of the heavens are uninhabited and that Earth of ours alone in peopled... we will suppose that in every region there are inhabitants...

Giordano Bruno would express similar ideas a century later. He believed the universe to contain an infinite number of worlds, that "there is not merely one world, one Earth, one sun, but as many worlds as we see bright lights around us" [6]. He also believed these worlds were inhabited by intelligent beings. Totally at odds with Christian beliefs, Bruno was imprisoned by the Inquisition in 1593, and kept in solitary confinement for seven years. Refusing to renounce his beliefs, he was burnt at the stake in 1600.

The idea of life on the Moon and planets in our solar system began to take root toward the end of the European Renaissance. Up until this time, Earth was thought to be the center of the universe with the sun, Moon, and stars moving in perfect circles around it. Although the observed motion of most stars did fit this model, that of certain 'wondering stars' (the planets) did not. This had led the Greek astronomer Aristarchus in the 2nd century B.C. to put the sun in the center, with Earth and the other planets revolving around it. Although Aristarchus' model did explain the retrograde motions of the planets (phenomenon in which the motion of a planet seems to stop and reverse direction), because he made the orbits circular instead of elliptical, the predictions of his model did not fit all of the observations exactly. So rather than abandon the Earth-centered view, it was modified by the introduction of epicycles. In this model, developed largely by Hipparchus but attributed to

Ptolemy in the 2nd century A.D., the planets moved in smaller circles (epicycles) that, in turn, moved in larger circles around the Earth.

In 1543, Nicholas Copernicus reversed the position of the sun and Earth, in effect, returning to the model proposed by Aristarchus 1700 years earlier. He did, however, keep the epicycles. In 1609, using Tycho Brahe's extensive observations of Mars, Johannes Kepler eliminated the epicycles and made the orbits elliptical. With Earth, now one of a handful of planets orbiting the sun, it had, on one hand, lost its special cosmological status, but on the other, opened the door to new speculation about life on the Moon and planets.

Early Speculation about Life on the Moon and Planets

Due to its proximity to Earth, early thought about life on other worlds involved the Moon. The Pythagoreans said "that the Earthy appearance of the moon is due to its being inhabited by animals and by plants, like those on our Earth, only greater and more beautiful..." [12]. In the first century A.D., the Graeco-Roman writer Plutarch discussed the habitability of the Moon in the form of a dialog between two characters, Theon and Lamprias. Theon, believing the Moon to be too hot and its atmosphere too tenuous to support life, asks why it exists at all since it serves "no purpose, neither bringing forth fruit nor providing for men of some kind of origin, an abode, and a means of life, the purposes for which this Earth of ours came into being" [9]. Lamprias responds that it might exist for reasons other than to sustain life, but may, given the diversities of Nature as evidenced on Earth, support life after all.

Fifteen hundred years later, inspired by Plutarch's work, Kepler would be the first to bring empirical evidence to bear in support of the idea that the Moon was a world like the Earth. He showed by means of an optical experiment that the lunar spots were not an illusion as Plutarch had claimed.

In 1609, using a telescope he had designed and constructed himself, Galileo observed the Moon close-up for the first time [13]:

> It is most beautiful and pleasing to the eye to look upon the lunar body, distant from us about sixty terrestrial diameters, from so near as if it were distant by only two of these measures, so that the diameter of the same Moon appears as if it were thirty times, the surface nine-hundred times, and the solid body about twenty seven thousand times larger than when observed only with the naked eye. Anyone will then understand with the certainty of the senses that the Moon is by no means endowed with a smooth and polished surface, but is rough and uneven and, just as the face of the Earth itself, crowded everywhere with vast prominences, deep chasms, and convolutions.

Commenting on the light and dark areas, he wrote: "if anyone wanted to resuscitate the old opinion of the Pythagoreans that the Moon is, as it were, another Earth, its brighter part would represent the land surface while its darker part would more appropriately represent the water surface." He also noted circular features, one of which appeared to be perfectly round, a feature Kepler found to be particularly interesting [13],

> I cannot help wondering about the meaning of that large circular cavity in what I usually call the left corner of the mouth [of the face of the Moon]. Is it a work of nature, or of a trained hand? Suppose there are living creatures on the Moon...

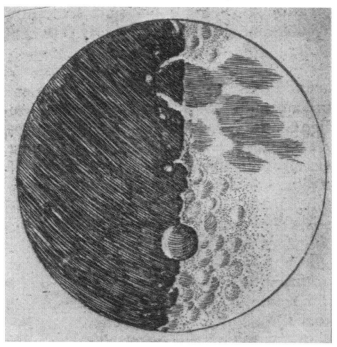

Figure 3 One of Galileo's 1609 drawings of the Moon from his treatise *Sidereus Nuncius* (1610).

In this letter to Galileo, Kepler speculates further [13]:

> It surely stands to reason that the inhabitants express the character of their dwelling place, which has much bigger mountains and valleys than our Earth has. Consequently, being endowed with very massive bodies, they also construct gigantic projects. Their day is as long as 15 of our days, and they feel insufferable heat. Perhaps they lack stone for erecting structures against the sun. On the other hand, maybe they have a soil as sticky as clay. Their usual building plan, accordingly, is as follows. Digging up huge fields, they carry out the Earth and heap it in a circle, perhaps for the purpose of drawing out the moisture down below. In this way they may hide in the deep shade behind their excavated mounds and, in keeping with the sun's motion, shift about inside, clinging to the shadow. They have, as it were, a sort of underground city. They make their homes in numerous caves hewn out of that circular embankment. They place their fields and pastures in the middle, to avoid being forced to go too far away from their farms in their flight from the sun.

Although Galileo found the Moon to be like Earth in many ways, unlike Kepler, he did not believe it was inhabited. A devote Catholic, Galileo, like his predecessors, had to maintain a delicate balance between scientific integrity and theological correctness. After all, Bruno had been executed by the Inquisition the decade before.

In 1640, John Wilkins, in *The Discovery of a World in the Moone*, discusses its similarity to the Earth, and the possibility that it might be inhabited. In it, he considers thirteen propositions [14]:

1. That the strangeness of this opinion is no sufficient reason why it should be rejected, because other certain truths have been formerly esteemed ridiculous, and great absurdities entertained by common consent;

2. That a plurality of worlds does not contradict any principle of reason or faith;

3. That the heavens do not consist of such pure matter which can privilege them from the like change and corruption, as these inferior bodies are liable unto;

4. That the Moon is a solid, compacted, opacous body;

5. That the Moon hath not any light of her own;

6. That there is a world in the Moon, hath been the direct opinion of many ancient, with some modern Mathematicians, and may probably be deduced from the tenants of others;

7. That those spots and brighter parts which by our sight may be distinguished in the Moon, do show the difference betwixt the Sea and Land in that other World;

8. That the spots represent the Sea, and the brighter parts the Land

9. That there are high mountains, deep valleys, and spacious plains in the body of the Moon;

10. That there is an atmosphere, or orb of gross vaporous air, immediately encompassing the body of the Moon;

11. That as their world is our Moon, so our world is their Moon;

12. That tis probable there may be such Meteors [wind, rain, dew, and moisture] belonging to that world in the Moon, as there are with us;

13. That tis probable there may be inhabitants in this other World, but of what kind they are is uncertain.

Wilkins used opinions expressed in ancient Greek and Roman texts together with observational data to argue some of these points, for the others, he reasoned by analogy, comparing features on the Moon to what he thought were similar features on Earth.

As the telescope continued to improve, that the planets were celestial bodies like Earth was becoming evident. In addition to viewing the moon through his telescope, Galileo, in 1610, observed the phases of Venus, moons of Jupiter, and faint structure around Saturn. The planet Mars, which Galileo could just barely resolve as a disk in 1610, was first drawn by Francesco Fontana in 1636. As the resolving power of the telescope improved, surface markings on Mars became visible. In 1659, Christiaan Huygens sketched the large feature on Mars we now know as Syrtis Major, and by measuring its movement, determined the rotational period of the planet to be 24 hours. In 1672, Huygens made a drawing of the planet that showed the south polar cap. Also during this period, Huygens discovered rings and moons around Saturn (1656), Robert Hooke observed the great red spot of Jupiter (1664), and Giovanni Cassini measured the rotational periods of Jupiter, Mars, and Venus (1665).

By the end of the 17th century, there was widespread belief in a plurality of worlds — that the planets were 'other Earths', and the stars were other suns, each with planets of their own. Most believed the planets to be inhabited. In 1688, Bernard de Fontenelle published a popular work entitled *Conversations on the Plurality of Worlds*. Fontenelle's imagination seems boundless as he imagines what life might be like on the other planets [15]:

> We, the Inhabitants of the Earth, are but one little Family of the Universe, we resemble one another. The Inhabitants of another Planet, are another Family, whose Faces have another Air peculiar to themselves; by all appearance, the difference increases with the distance, for could one see an Inhabitant of the Earth, and one of the Moon together, he would perceive less difference between them, than between an Inhabitant of the Earth, and an Inhabitant of Saturn. Here (for example) we have the use of the Tongue and Voice, and in another Planet, it may be, they only speak by Signs. In another the Inhabitants speaks not at all. Here our Reason is formed and made perfect by Experience. In another Place, Experience adds little or nothing to Reason. Further off, the old know no more than the young. Here we trouble our selves more to know what's to come, than to know what's past. In another Planet, they neither afflict themselves with the one nor the other; and 'tis likely they are not the less happy for that. Some say we want a sixth Sense by which we shou'd know a great many things we are now ignorant of. It may be the Inhabitants of some other Planet have this advantage; but want some of those other five we enjoy; it may be also that there are a great many more natural Senses in other Worlds; but we are satisfied with the five that are fallen to our Share, because we know no better.

Toward a Scientific Theory of Extraterrestrial Life

Continued refinement of the telescope led to an explosion of astronomical knowledge in the seventeenth and eighteenth centuries. Early in the 18th century, Cassini's nephew, Giacomo Maraldi, observed whitish spots near both of Mars' poles. He also detected changes in the South Pole, and noticed what appeared to be variable markings on Mars. In 1783, Sir William Herschel made extensive observations of the south pole of Mars. From his measurements he estimated the inclination of its axis to the plane of its orbit to be approximately 28 degrees. On Earth, it is the inclination of our axis (about 25 degrees) that gives rise to the seasons. This similarity between Earth and Mars suggested to Herschel that the whitish spots in the Polar Regions were made of snow and ice, and that observed changes in the polar cap were caused by the seasonal melting of snow and ice by the sun.

Meanwhile, in his treatise, *Universal Natural History and Theory of Heaven*, Immanuel Kant proposed a new theory to explain the formation of the solar system, which at the time (1755) consisted of six planets: Mercury, Venus, Earth, Mars, Jupiter and Saturn. A century earlier, Rene Descartes had proposed a mechanical model of the universe based on particles and vortices. Kant's theory, which was subsequently refined by Pierre Simon Marquis de Laplace, became known as the Nebular Theory, and is still accepted with certain modifications today. According to this theory, the solar system began as a large cloud of dust (nebula), which was slowly spinning. Gradually, the force of gravity caused the cloud to collapse. As it collapsed, by the law of conservation of angular momentum, its spin increased, causing the cloud to flatten into a pancake with a central bulge. Within the pancake, variations in density caused local regions to collapse in a similar way. As the central bulge and surrounding regions continued to collapse, they began to heat up. The central bulge, by virtue of its size, became the sun. The surrounding regions eventually cooled to become the planets and moons.

Toward the end of his treatise, Kant uses his theory as a point of departure for speculating on whether or not the planets are inhabited: "I believe that it is not necessary to assert that all planets must be inhabited... Perhaps all the celestial bodies have not yet completely developed" [16]. At this time, scholars believed the Earth to be about six thousand years old. Given this, Kant states, "Hundreds and maybe thousands of years are necessary for a large celestial body to reach a stable material condition... Earth perhaps existed for a thousand years or more before it was in a condition to support human beings, animals, and plants." He goes on to discuss the physical characteristics and spiritual capacities of living creatures on other planets according to their distance from the sun

> Human nature, which in the scale of being holds, as it were, the middle rung, is located between two absolute outer limits, equidistant from both. If the idea of the most sublime classes of reasoning creatures living on Jupiter or Saturn makes human beings jealous and discourages them with the knowledge of their own humble position, a glance at the lower stages brings content and calms them again. The beings on the planets Venus and Mercury are far below the perfection of human nature.

Around the same time, Georges Buffon proposed another theory for the formation of the solar system — that the planets were formed from material ripped from the Sun by a collision with a comet. According to Buffon's theory it would take the Earth about 75,000 years to cool.

Subsequent discoveries in the terrestrial sciences led to greatly increased estimates of the age of the Earth, from tens of thousands, to hundreds of millions, and finally within the last century, to billions of years. Two theories were being debated at the start of the 19th century: catastrophism — that the Earth has been shaped by a series of great catastrophes, and uniformitarism — that the same geological processes that cause changes in the Earth's surface today have been have operating the same way over very long periods of time. Georges Cuvier, an early catastrophist, believed the world has undergone several upheavals in its history, the last occurring five or six thousand years ago. James Hutton, and later Charles Lyell, the father of uniformitarianism, believed the Earth to be much older. In 1860, it was calculated to be about 3 million years old based on the thickness of sedimentary rock. Other methods yielded estimates up to several hundred million years. Today, it is widely believed that our planet formed about 5 billion years ago.

Growing scientific consensus that the Earth has been around for a very long time set the stage, in 1859, for Charles Darwin's theory of evolution. Darwin's theory holds that biological organisms exist in permanent states of change, and that all life struggles to exist. Through a process known as natural selection, organisms that are well-adapted to the environment in which they live survive and flourish while those that are not eventually die out. The changes or variations that ultimately result in increased survivability are random in nature and occur over enormously long periods of time.

The extraterrestrial implication of this theory was obvious: wherever life started, it was likely that given enough time, it would produce intelligent beings in the end. Together with Laplace and Kant's nebular theory, Darwin's theory of evolution would provide the first truly scientific basis for considering the possibility of extraterrestrial life — one that would be applied in just a few years to the planet Mars.

The Canals on Mars

Around the middle of the 19th century, astronomers began to notice the colors and tones of Mars. Johann Madler and his colleague Wilhelm Beer studied the darker regions, and noticed changes in the area surrounding the north polar cap. This suggested the possibility that the dark area might be marshy soil, wet from water released by the melting ice cap.

In 1858, a Catholic priest named Angelo Secchi suggested the darker areas might be seas separated by lighter-colored continents. He went as far as to state that changes in the poles were proof that water and seas exist on Mars. In the fall of 1877, during periods of good 'seeing' with his telescope, Giovanni Schiaparelli drew extremely detailed maps of Mars. From time to time, he noticed what appeared to be linear features. Believing the dark areas on Mars to be seas, he reasoned that the lines were channels or rivers, which he called *canali*. Schiaparelli believed what he saw was real, but never doubted the *canali* were natural. Because of his rather schematic style of drawing, the features he drew took on an artificial character. This led Camille Flammarion to make the following remarks in 1892 [17]:

> On a globe could Nature trace such straight lines, cutting each other in such a fashion?... The more we look at these drawings, the less that we can attribute them to blind chance... The actual conditions on Mars are such that it would be wrong to deny that it could be inhabited by human species whose intelligence and methods of action could be far superior to our own...

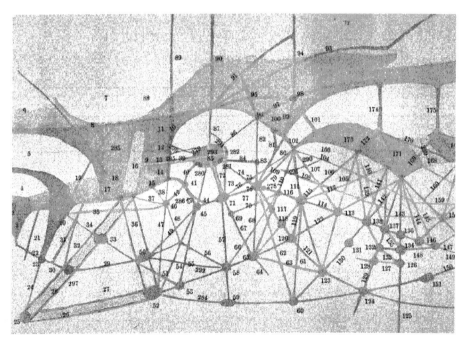

Figure 4 A portion of an early map of Mars drawn by Percival Lowell showing the canals. (Lowell Observatory)

Shortly thereafter, Percival Lowell became interested in Schiaparelli's work. From the start, he took the *canali* to be artificially-constructed canals or waterways. At the time, a popular interpretation of the Kant-Laplace theory was that planets farthest from the Sun had cooled and solidified first. Since Mars was farther from the sun than Earth, it was older. And because it was older, evolution had progressed further than on Earth. Applying these ideas, Lowell believed that "Whatever the particular planet's line of development,... it proceeds to greater and greater degrees of evolution..." [18]. With respect to

Mars, "we have before us the spectacle of a world relatively well on in years, a world much older than the Earth... His continents are all smoothed down; his oceans have all dried up..." And "Mars being thus old himself, we know that evolution on his surface must be similarly advanced." In referring to the canals,

> The evidence of handicraft, if such it be, points to a highly intelligent mind behind it... Certainly what we see hints at the existence of beings who are in advance of, not behind us, in the journey of life.

Lowell believed the Martians had surpassed us technologically, and, in an attempt to save their dying world, had created a network of canals to distribute water from the polar caps over the planet. Consistent with the implications of Darwinian evolution, he concludes by saying,

> To talk of Martian beings is not to mean Martian men... Even on this Earth man is of the nature of an accident... Amid the surroundings that exist on Mars, surroundings so different from our own, we may be practically sure other organisms have been evolved of which we have no cognizance. What manner of beings they may be we lack the data even to conceive.

While Lowell was advancing these ideas, astronomers were beginning to doubt the reality of the Martian canals. More detailed observations by Edward Barnard in 1894 led him to conclude that what Schiaparelli and others saw as linear features were illusionary. This conclusion was later confirmed by a resolution study performed by Edward Maunder in which he demonstrated the eye's tendency to see poorly resolved patterns of dots as linear features. Last, but not least, spectroscopic measurements showed Mars to be a colder, drier and less habitable planet than had been thought. The Earth-like picture of Mars, which had developed over the past century, was beginning to change. As the 19th century drew to a close, so too did the idea that there was life on Mars.

Sources

1. *King James Bible* (http://quod.lib.umich.edu/k/kjv/).
2. Zecharia Sitchin, *The 12th Planet*, Avon Books, New York, 1976.
3. Erich von Daniken, *Chariots of the Gods?*, Bantam Books, New York, 1971.
4. Carl Sagan, "Direct Contact Among Galactic Civilizations by Relativistic Interstellar Flight," *Planetary and Space Science*, Vol. 11, pp 485-498, 1963.
5. N.M. Swerdlow, *The Babylonian Theory of the Planets*, Princeton University Press, Princeton, New Jersey, 1998.
6. John Burnet, *Early Greek Philosophy*, (http://www.classicpersuasion.org/pw/burnet/index.htm).
7. Epicurus, *Letter to Pythocles*, 06-270 B.C.E (http://www.grtbooks.com).
8. Plato, *Timaeus*, 360 B.C.E (http://classics.mit.edu/Plato/timaeus.html).

9. Steven J. Dick, *Plurality of Worlds, The Origins of the Extraterrestrial Life Debate from Democritus to Kant*, Cambridge University Press, Cambridge, 1981.
10. Aristotle, *On the Heavens* (Book 1), Written 350 B.C.E, (http://classics.mit.edu/Aristotle/heavens.html).
11. Lucretius, *On the Nature of Things*, 50 B.C.E (http://classics.mit.edu/Carus/nature_things.html).
12. Arthur Fairbanks, *The First Philosophers of Greece* (http://history.hanover.edu/texts/presoc/pythagor.html).
13. Galileo Galilei, *Siderius Nuncius*, Translated by Albert Van Helden, The University of Chicago Press, Chicago, 1989.
14. John Wilkins, *The Discovery of a World in the Moone*, 1640 (http://www.gutenberg.org/etext/19103).
15. Aphra Behn, *A Discovery of New Worlds*, 1688.
16. Immanuel Kant, *Universal Natural History and Theory of Heaven*, 1755 (http://www.mala.bc.ca/~johnstoi/kant/kant2e.htm).
17. William Sheehan, *The Planet Mars — A History of Observation and Discovery*, University of Arizona Press, Tucson, 1996.
18. Percival Lowell, *Mars*, 1895 (http://www.bibliomania.com/NonFiction/Lowell/Mars/index.html).

Three — Terrestrial Intelligence

> Lowell always said the regularity of the canals was an unmistakable sign that they were of intelligent origin. This is certainly true. The only unresolved question was which side of the telescope the intelligence was on. — Carl Sagan

In the aftermath of the Martian canal controversy, serious scientific speculation concerning the possibility of extraterrestrial life began to fade. Works of science fiction by Jules Verne and H.G. Wells appeared around this time. Picking up from where Lowell had left off, H.G. Wells published *The War of the Worlds* in 1898. Echoing Lowell's theories about Mars and its inhabitants [1]:

> It must be, if the nebular hypothesis has any truth, older than our world; and long before this Earth ceased to be molten, life upon its surface must have begun its course…. Nor was it generally understood that since Mars is older than our Earth, with scarcely a quarter of the superficial area and remoter from the sun, it necessarily follows that it is not only more distant from time's beginning but nearer its end.

Lowell's Martians were, by virtue of their age, wise and benevolent; however, Wells conceived of them differently:

> No one gave a thought to the older worlds of space as sources of human danger, or thought of them only to dismiss the idea of life upon them as impossible or improbable… It was the beginning of the rout of civilization, of the massacre of mankind.

The War of the Worlds was the first novel about what actual contact between two worlds might be like. It was as much a social commentary as science fiction with the violence of the Martian invasion foreshadowing the horror and destruction of World War I, later referred to by H.G. Wells as "the war to end war."

Forty years later, on Halloween night in 1938, Orson Welles' Mercury Theatre dramatization of *The War of the Worlds* was heard by almost two million radio listeners. Some, who tuned into the broadcast, did not know they were listening to a dramatic recreation of H.G. Wells' book. Instead, on hearing the special bulletins and sound effects, they actually thought Earth was being invaded by Martians.

Then, less than a decade later, life would seemingly imitate art.

Unidentified Flying Objects

During World War II, military pilots began to report glowing balls of light that would fly along their aircraft and maneuver rapidly. On 24 June, 1947, a civilian pilot, flying over the Cascade Mountains in Washington, observed nine flying disc-shaped aircraft, traveling in formation at a high rate of speed. The pilot was Kenneth Arnold. His description of the movement of these objects, "like a saucer would if it was skipped across the water," led to the popular term 'flying saucer'. Although there is evidence that similar objects had been seen before, this report was the first to gain widespread attention in the media.

A few weeks later, the Army Air Force reported the recovery of a flying saucer in the New Mexico desert, near the town of Roswell. The following article appeared in the Roswell Daily Record on July 8, 1947:

> The intelligence office of the 509th Bombardment group at Roswell Army Air Field announced at noon today, that the field has come into possession of a flying saucer. According to information released by the department, over authority of Maj. J. A. Marcel, intelligence officer, the disk was recovered on a ranch in the Roswell vicinity, after an unidentified rancher had notified Sheriff Geo. Wilcox, here, that he had found the instrument on his premises. Major Marcel and a detail from his department went to the ranch and recovered the disk, it was stated. After the intelligence officer here had inspected the instrument it was flown to higher headquarters. The intelligence office stated that no details of the saucer's construction or its appearance had been revealed...

But the next day, Major Marcel's superiors said they had made a mistake — the wreckage was that of a weather balloon. One version of what might have happened is given in the controversial MJ-12 briefing document, which was supposedly prepared for president-elect Dwight Eisenhower [2]:

> On 7 July, 1947, a secret operation was begun to assure recovery of the wreckage of this object for scientific study. During the course of this operation, aerial reconnaissance discovered that four small human-like beings had apparently ejected from the craft at some point before it exploded. These had fallen to Earth about two miles east of the wreckage site. All four were dead and badly decomposed due to action by predators and exposure to the elements during the approximately one week time period which had elapsed before their discovery. A special scientific team took charge of removing these bodies for study... The wreckage of the craft was also removed to several different locations... Civilian and military witnesses in the area were debriefed, and news reporters were given the effective cover story that the object had been a misguided weather research balloon.

The MJ-12 document, which appeared in the late 1980's, is a subject of considerable debate. Although skeptics have labeled it a hoax, UFO researcher Stanton Friedman has produced government memoranda that appear to corroborate the existence of MJ-12. Wilbert Smith, in a formerly classified Canadian government memorandum dated November 21, 1950 states, "The matter is the most highly classified subject in the United States Government, rating higher even than the H-bomb. Flying saucers exist. Their modus operandi is unknown but a concentrated effort is being made by a small group headed by Dr. Vannevar Bush" [3].

Meanwhile UFO sightings continued through the 1950's. Again, from the Eisenhower briefing document:

> Hundreds of reports of sightings of similar objects followed. Many of these came from highly credible military and civilian sources. These reports resulted in independent efforts by several different elements of the military to ascertain the nature and purpose of these objects in the interests of national defense. A number of witnesses were interviewed and there were several unsuccessful attempts to utilize aircraft in efforts to pursue reported discs in flight.

A series of dramatic sightings occurred over Washington, DC, late in July 1952. Seen by pilots as well as thousands on the ground, the objects were tracked by multiple radars. On two occasions, Air Force jets were even dispatched to the scene. Yet, despite objections from an experienced air controller, the official explanation given was that the sightings were caused by a temperature inversion — a meteorological phenomenon that can cause 'ghost' blips to appear on a radar screen.

That the government, in particular the Central Intelligence Agency, was attempting to debunk UFOs at the time has been revealed in a classified report (since declassified) authored in February 1953 by a scientific advisory panel headed by H. P. Robertson, a professor at the California Institute of Technology (Caltech). Although the panel believed UFOs posed no direct threat to national security, they felt there was a potential danger that too many reports could overload official channels, so much so, that the Air Force might not be able to react to a 'real' threat. The panel recommended that military personnel be better trained in analyzing reported sightings, and that a program of 'debunking' be instituted to reduce public interest in flying saucers [4]

> The 'debunking' aim would result in reduction of public interest in 'flying saucers' which today evokes a strong psychological reaction. This education would be accomplished through mass media such as television, motion pictures, and popular articles. Basis of such education would be actual case histories which had been puzzling at first but later explained. As in the case of conjuring tricks, there is much less stimulation if the 'secret' is known. Such a program should tend to reduce the current gullibility of the public and consequently their susceptibility to clever hostile propaganda...
>
> Members of the Panel had various suggestions related to the planning of such an educational program. It was felt strongly that psychologists familiar with mass psychology should advise on the nature and extent of the program. In this connection, Dr. Hadley Cantril (Princeton University) was suggested. Cantril authored 'Invasion from Mars', (a study in the psychology of panic, written about the famous Orson Welles radio broadcast in 1938) and has since performed advanced laboratory studies in the field of perception... Also, someone familiar with mass communications techniques, perhaps an advertising expert, would be useful. Arthur Godfrey [a popular radio personality at the time] was mentioned as possibly a valuable channel of communication reaching a mass audience of certain levels...

The Jim Handy Co. which made World War II training films (motion picture and slide strips) was also suggested as well as Walt Disney, Inc. animated cartoons. [It was also] suggested that the amateur astronomers in the U.S. might be a potential source of enthusiastic talent 'to spread the gospel'....

It was also noted that although the panel did not believe UFOs to be extraterrestrial in origin, they did accept the possibility that Earth might some day be visited by extraterrestrial beings. Most of the group, with the exception of Robertson, felt that the existence of extraterrestrial artifacts posed no threat and should be subject to scientific study. Robertson, on the other hand, felt that "such artifacts would be of immediate and great concern not only to the U.S. but to all countries" and would serve to unite the world again a common threat.

The Search for Extraterrestrial Intelligence

While the UFO phenomenon was unfolding in the 1950's, radio astronomers were considering the possibility of using radio telescopes to search for extraterrestrial intelligence. However, the idea of extraterrestrial communication was not new.

Early in the 19th century, the great German mathematician, Karl Gauss proposed the universal language of mathematics, in particular the Pythagorean theorem (that the square of the hypotenuse of a right triangle is equal to the sum of the squares of the other two sides), be used to communicate with the planets [5]. The idea was to plant tracts of pine forest in Siberia to enclose a huge right triangle with wheat growing inside. In summer, the light-colored wheat would strike a contrast to the dark green of the trees, while in winter, snow inside would stand out from the trees. A similar idea, suggested by Joseph Johann von Littrow, was to dig a series of geometrically-shaped trenches in the Sahara desert 20 miles in size, and on successive nights, to fill one at a time with kerosene and light it. Charles Cros of France proposed that a huge mirror be built to focus sunlight and burn out messages on the desert sands of Mars. The German astronomer Joseph Plassmann thought about visual communication from the other perspective and wondered if Martians could see the lights of our cities.

Toward the end of the 19th century, developments in electromagnetic theory led to the first demonstration of radio waves by Heinrich Hertz in 1888. Radio was developed shortly thereafter, first by Nikola Tesla in 1893, and then by Guglielmo Marconi a few years later. According to Tesla, one night in 1899 while working on a very sensitive radio receiver in his Colorado Springs laboratory, he detected signals that occurred

> periodically, and with such a clear suggestion of number and order that they were not traceable to any cause then known to me. I was familiar ... with such electrical disturbances as are produced by the sun, Aurora Borealis and Earth currents, and I was as sure as I could be of any fact that these variations were due to none of these

causes. The nature of my experiments precluded the possibility of the changes being produced by atmospheric disturbances ... Although I could not decipher their meaning, it was impossible for me to think of them as having been entirely accidental ... a purpose was behind these signals ...they are the results of an attempt by some human beings, not of our world, to speak to us by signals ... I am absolutely certain that they are not caused by anything terrestrial [6].

Knowing that he would be ridiculed by fellow scientists, Tesla was reluctant to reveal his discovery. When he finally did, criticism from his fellow scientists was predictable:

Mr. Nikola Tesla has announced that he is confident that certain disturbances of his apparatus are electrical signals received from a source beyond the Earth. They do not come from the sun, he says; hence they must be of planetary origin, he thinks; probably from Mars, he guesses. It is the rule of sound philosophizing to examine all probable causes for an unexplained phenomenon before invoking improbable ones...

Eventually, other scientists followed Tesla's lead. In 1921, while conducting atmospheric experiments in the Mediterranean, Marconi claimed to have detected regular pulses, which he believed originated "at some point in outer space." Even though television had not yet been invented, it was Marconi's expectation that interplanetary communication would be a "transmission of pictures accompanied by a simple code" [6].

In August 1924, as Mars approached to within 35 million miles of Earth, David Todd, with the help of the U.S. Government, arranged for all countries in the world with high-power transmitters to turn off their equipment for 5 minutes every hour for a period of several days. During that time, incoming radio signals were converted to light and printed on a reel of photosensitive tape five inches wide. The printed tape, consisting of 16 printed frames, was 25 feet long, and contained patterns of dots and dashes, some of which appeared in groups spaced about 30 minutes apart.

It is likely that what Tesla and the others heard was extraterrestrial radio noise. In 1933, Karl Jansky discovered its source at the center of the Milky Way galaxy, in the constellation of Sagittarius. This led to the beginning of modern radio astronomy a few years later.

The era of modern SETI, the search for extraterrestrial intelligence, began in 1959 when Giuseppi Cocconi and Philip Morrison proposed that radio telescopes be used to search for intelligent signals in the microwave region of the electromagnetic spectrum. The following year, Frank Drake conducted the first microwave search (Project Ozma) for extraterrestrial radio signals at the National Radio Astronomy Observatory in Green Bank, West Virginia. Other projects with names like Cyclops, META, and SERENDIP followed.

Radio SETI assumes that: 1) there are advanced technological civilizations in the galaxy either communicating with one another or trying to communicate with us, 2) radio, specifically microwave radiation near the 21 cm wavelength of hydrogen, is their preferred means of communication, 3) our radio telescopes are powerful enough to detect their signals, and 4) we will be able to understand the message.

All four assumptions are open to question. At one extreme, some say that we are the only advanced civilization in the galaxy, since if there were others, we would know about them. At the other extreme, some have estimated upwards of billions of advanced technological civilizations in the galaxy. The scientific consensus is that the number of advanced civilizations is somewhere in between the two extremes, and is sufficient to warrant some kind of search. Yet, from Tesla's experiments around the turn of the century to today's SETI Institute, extraterrestrial radio signals of intelligent origin have yet to be detected.

Over more than a century, continued advances in electrical engineering and signal processing have led to increasingly more refined searches both in terms of space and frequency, with smaller and smaller regions of space are being examined at narrower and narrower frequencies. So why the lack of success? Have we not looked in the right directions, or listened in at the correct frequencies? One explanation suggested by James Deardorff is that Earth is under a kind of cosmic embargo [7]. His theory is that

> our Galaxy is nearly saturated with extraterrestrial life forms, that our existence requires in hindsight that they were and are benevolent toward us [or at least need us in some way], and that our lack of detection of them or communications from them implies that an embargo is established against us to prevent any premature knowledge of them.

Deardorff suggests the embargo "must be a leaky one designed to allow a gradual disclosure of the alien message and its gradual acceptance on the part of the general public over a very long time-scale." One way to achieve this would be to communicate in a way that is easily "accessible to the general public but in a form not acceptable or believable to scientists. Government agencies, upon advice from scientists, would then take no actions, and the embargo would more or less remain intact."

Alternative SETI Concepts

A way of increasing the likelihood of success in SETI is to narrow the search — to focus on a smaller portion of the galaxy, say our own solar system. Another way is to explore other means of contact.

In 1963, Carl Sagan published a paper entitled, "Direct Contact Among Galactic Civilizations by Relativistic Interstellar Flight." Using Frank Drake's famous equation for estimating the number of advanced technological civilizations in the galaxy, and assuming the galaxy contains a "loosely integrated community of diverse civilizations cooperating in the exploration and sampling of astronomical objects and their inhabitants," Sagan estimated that a planetary system with intelligent life could be visited once every 10,000 years or so [8]. He goes on to say:

> It follows that there is the statistical likelihood that Earth was visited by an advanced extraterrestrial civilization at least once during historical times. There are serious difficulties in demonstrating such a contact by ancient writing and iconography alone. Nevertheless there are legends which might profitably be studied in this context.

He then mentions the Babylonian account of the mysterious origin of the Sumerian civilization (discussed in Chapter 2), and concludes by speculating that: "Bases or other artifacts of interstellar spacefaring civilizations might also exist elsewhere in the solar system." That same year, in a speech to the American Rocket Society, Sagan expanded further on this idea:

> Because of weathering and the possibility of detection and interference by the inhabitants of Earth, it would be preferable not to erect such a base on the Earth's surface. The moon seems one reasonable alternative. Forthcoming photographic reconnaissance of the moon from space vehicles, particularly of the back, might bear these possibilities in mind.

Photoreconnaissance of the Moon

For hundreds of years, astronomers have noticed strange features on the lunar surface. The first recorded sighting was a star-like point of light on the dark side in 1540. Other features that have been reported include small whitish clouds by Cassini in 1671, 'lightning' by Louville and Halley in 1715, 'vapors' by Schroter and Olber in 1797, brilliant flashing spots on the dark side by Gruithuisen in 1821, dots and streaks of light by Slack and Ingall in 1869, a glow of light in the crater Plato by Fauth in 1907, a moving luminous speck near Gassendi by Haas in 1941, a pulsating spot on dark side by Emanuel and others in 1965, and hundreds more. In 1968, NASA published a compendium of these so-called lunar transient phenomena[2] (LTP) sightings in a report entitled, "Chronological Catalog of Reported Lunar Events."

Several years after Sagan's comments on the possibility of finding alien artifacts on the Moon, a Lunar Orbiter spacecraft photographed an unusual collection of objects (Figure 5) on the western edge of the Sea of Tranquility [9]. William Blair, an anthropologist at Boeing, the company that made the Lunar Orbiter, was reminded of patterns seen in aerial survey maps of prehistoric archaeological sites. Intrigued by the length of their shadows, apparent

---

[2] See http://en.wikipedia.org/wiki/Transient_lunar_phenomenon for more information on LTPs.

geometrical layout, and proximity to an unusual rectangular depression, Blair believed the objects might be artificial:

> If such a complex of structures were photographed on Earth, the archaeologist's first order of business would be to inspect and excavate test trenches and thus validate whether the prospective site has archaeological significance [10].

Richard Shorthill, a geologist at Boeing disagreed: "There are many of these rocks on the moon's surface. Pick some at random and you eventually find a group that seems to conform to some kind of pattern."

Figure 5 Lunar Orbiter photograph LO2-61H3 containing a number of objects casting long shadows (top middle, and left). The line down the middle is a seam between two photographs. White crosses are registration marks on the camera and are spaced about 750 feet apart. In this photo, north is to the left. (NASA/Lunascan)

This was not the first time someone claimed to have discovered artificial objects on the Moon. In 1822, the astronomer Franz von Paula Gruithuisen announced that he saw a walled city on the moon (Figure 6) through his telescope, located near the crater Schroter [5]. Although the moon had been a subject of considerable speculation concerning extraterrestrial life for thousands of years, this was the first time anyone claimed to have found physical evidence of it. Subsequent observations by other astronomers ultimately revealed Gruithuisen's 'city' to be nothing more than a chance alignment of craters and other features on the lunar surface (Figure 7).

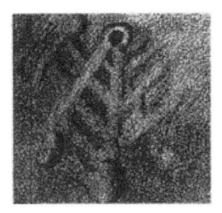

Figure 6 Gruithuisen's sketch of an artificial structure which he claimed to have seen on the Moon near the crater Schroter.

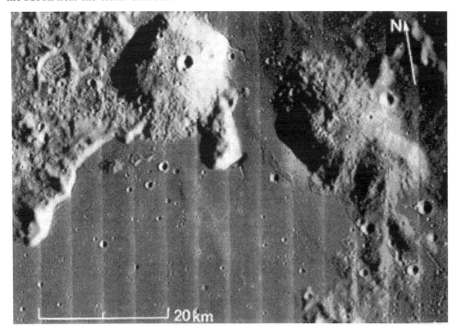

Figure 7 Lunar Obiter photograph of the Moon. The feature Gruithuisen found is at the lower right. (NASA)

As Lunar Orbiter and the Apollo astronauts continued to photograph the Moon, other anomalies were captured on film. George Leonard discusses some of these features in *Somebody Else is On the Moon*, published in 1976. He describes a variety of strange features including geometrical shapes (linear features, parallel lines, crosses, and others), domes, and what he believes to be bridges, construction activities, pieces of machinery, and vehicles. Although the photographic evidence is not particularly compelling, Leonard does attempt to catalog and document anomalies on the lunar surface in much the same way NASA had cataloged LTPs in 1968. A few years later in 1981, Fred Steckling published additional photographic evidence of lunar anomalies in *We Discovered Alien Bases on the Moon.*

In June of 1985, Carl Sagan wrote an article for *Parade* magazine in which he comments on these features [13]:

> Around the time of the Apollo lunar landings, a number of people — amateur astronomers, flying saucer zealots, and writers for aerospace magazines — poured over the photographs searching for anomalies that NASA scientists and astronauts had overlooked. There were reports of enormous Latin letters and Arabic numerals inscribed on the lunar surface, pyramids, highways, crosses, glowing UFOs and, it was asserted, the long shadows of ballistic missiles, probably Soviet, aimed at Earth. All have turned out to be natural lunar geological formations misunderstood by the analysts, internal reflections in hand held cameras, and the like.

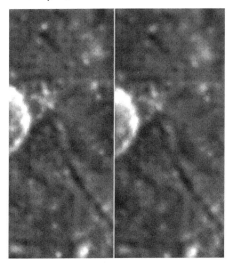

Figure 8 Stereo view[3] of a cross or x-shaped pattern next to a crater. This region is west-southwest of the crater Endymion on the near side of the moon. These views were constructed from Clementine images LUA4282n and 4315n.283. (NASA)

One might be tempted to defer to Sagan's authority on the matter if it were not for the fact that some of these anomalies have also been seen in more recent imagery. After a two decade-long lapse in lunar exploration, NASA and the Department of Defense placed the Clementine spacecraft into orbit around the Moon in the spring of 1994. Cameras aboard the craft collected over one million images of the moon at an average resolution of 400 meters per pixel. After examining several orbits worth of data, I was able to find two images containing the cross or 'X'-shaped pattern, found by Leonard and Steckling, one near the crater Endymion, and the other next to Vitello. Unlike Lunar Orbiter's film camera, Clementine's digital cameras captured one image after another as the spacecraft passed over the lunar surface. By using consecutive images taken a few seconds apart, stereo images can be created. In stereo, the 3-D shape of an 'X' found near the crater Endymion is particularly striking (Figure 8).

---

[3] To see this image in 3-D start by holding the book level and at arm's length and looking just over the top of the book at a distant object. Without actually lowering your gaze onto the page notice that the image pairs now seem to overlap to form a third image between them. Keeping your eyes relaxed and still focused in the distance, slowly and carefully shift your attention (but not your eyes) to that third, middle image. After some practice you should be able to focus clearly on the stereo image and then to draw the book toward you for a closer look.

In 1998, Alexey Arkhipov published a paper in the *Journal of the British Interplanetary Society* entitled "Earth-Moon system as a collector of alien artifacts" in which he describes an unusual formation near the crater Lovelace [14]:

> The formation looks like an isolated quasi-rectangular cluster of rectangular depressions (the collapse of some subsurface caves?)... The rectangularity and regularity of this ruin-like pattern is similar to modern projects for the lunar base as a subsurface construction protected from meteoroids and radiation.

Figure 9 Image LO2-61H3 rotated 50 degrees clockwise. With the sides of the rectangular depression aligned horizontally and vertically, one can see subtle rectilinear depressions and patterns above and to the left of the trench. (NASA)

 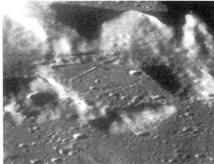

Figure 10 Rectangular features are not uncommon on the Moon[4]. These photos from Lunar Orbiter (left) and Apollo 10 (right) are of a rectangular region southeast of the crater Ukert. The orientation of this feature is roughly the same as the one near the Blair spires. (NASA)

---

[4] Many features on the Moon are oriented in northeast and northwest directions. This arrangement produces a global, rectilinear pattern known as the 'lunar grid', the origin of which is unknown [15].

To date, no known geological process has been suggested to explain the existence of these and other unusual features on the Moon. Yet Sagan, who suggested the possibility of finding extraterrestrial artifacts on the lunar surface in the first place, suddenly seemed to change his mind when confronted with evidence that possible artifacts might actually exist there. Why?

The Brookings Report

In April 1961, a report prepared for NASA by the Brookings Institution, *Proposed Studies of Peaceful Space Activities for Human Affairs*, was submitted to Congress. The purpose of the report was to identify the long-range implications for American society of space exploration. What is of particular interest, were Brookings' thoughts about the implications of the discovery of extraterrestrial life [16]:

> Recent publicity given to efforts to detect extraterrestrial messages via radio telescope has popularized — and legitimized — speculations about the impact of such a discovery on human values. It is conceivable that there is semi-intelligent life in some part of our solar system or highly intelligent life which is not technologically oriented, and many cosmologists and astronomers think it very likely that there is intelligent life in other solar systems. While face-to-face meetings with it will not occur within the next 20 years (unless its technology is more advanced than ours, qualifying it to visit Earth), artifacts left at some point in time by these life forms might possibly be discovered through our space activities on the Moon, Mars, or Venus...

In a way, echoing Robertson's concern about the discovery of extraterrestrial artifacts back in 1953, the Brookings report goes on to say "The knowledge that life existed in other parts of the universe might lead to a greater unity of men on Earth, based on the 'oneness' of man or on the age-old assumption that any stranger is threatening." Then, pointing out the possibility of societal disintegration resulting from contact with extraterrestrials, the report says this: "Anthropological files contain many examples of societies sure of their place in the universe, which have disintegrated when they had to associate with previous unfamiliar societies espousing different ideas and different life ways; others that survived such an experience usually did so by paying the price of changes in values and attitudes and behavior." Concluding this part of the report, two research areas were recommended:

> Continuing studies to determine emotional and intellectual understanding and attitudes — and successive alterations of them if any — regarding the possibility and consequences of discovering intelligent extraterrestrial life.

and

> Historical and empirical studies of the behavior of peoples and their leaders when confronted with dramatic and unfamiliar events and social pressures. Such studies

might help to provide programs for meeting and adjusting to the implications of such a discovery. Questions one might wish to answer by such studies would include: How might such information, under what circumstances, be presented to or withheld from the public for what ends. What might be the role of the discovering scientists and other decision makers regarding release of the fact of discovery?

Scientific Study of UFOs

Starting in 1947, the Air Force investigated UFOs under Project Blue Book. Up until the project was terminated in 1969, a total of 12,618 sightings were reported. Although 701 of the sightings were labeled 'unidentified', the Air Force's position was that none of them indicated technological developments or principles beyond the range of current scientific knowledge, and that there was no evidence indicating the sightings were extraterrestrial in origin.

The decision to discontinue UFO investigations followed a 1968 report entitled, *Scientific Study of Unidentified Flying Objects*, headed by Edward U. Condon from the University of Colorado. The main conclusion of the report was "that nothing has come from the study of UFOs in the past 21 years [since 1947] that has added to scientific knowledge. Careful consideration of the record as it is available to us leads us to conclude that further study of UFOs probably cannot be justified in the expectation that sciences will be advanced thereby" [17].

They explain their conclusion in this way:

> It has been argued that this lack of contribution to science is due to the fact that very little scientific effort has been put on the subject. We do not agree. We feel that the reason there has been very little scientific study of the subject is that those scientists who are most directly concerned, astronomers, atmospheric physicists, chemists, and psychologists, having had ample time to look into the matter, have individually decided that UFO phenomena do not offer a fruitful field in which to look for major scientific discoveries.

To say that a UFO sighting is 'unidentified' is to say that it is anomalous. Many important scientific discoveries have resulted from the study of anomalies. But if so little effort has been spent studying UFOs, how can so many scientists have come to such a strong conclusion — that no further study is needed? Scientific breakthroughs require years of hard work. Why would the study of UFOs be any different? They go on to say that:

> Although we conclude after nearly two years of intensive study, that we do not see any fruitful lines of advance from the study of UFO reports, we believe that any scientist with adequate training and credentials who does come up with a clearly defined, specific proposal for study should be supported.

In the report, it is stated that the parallel channels of funding that exist within the federal government (at the time) would prevent the suppression of important new information pertaining to UFOs, should it become available.

Instead of viewing the UFO phenomena as a mystery to be solved, Condon considered it a 'problem' to be dealt with. The report labels supporters of UFOs irresponsible, and strongly recommends that educators discourage students from reading about UFOs. Instead, they should be encouraged to pursue "serious study of astronomy and meteorology, and in the direction of critical analysis of arguments for fantastic propositions that are being supported by appeals to fallacious reasoning or false data." In other words, any data supporting the existence of UFOs is, by definition, false.

UFOs in Space

As U.S. astronauts and Soviet cosmonauts ventured into space during the 1960's, UFOs seemed to follow. Even the Condon report admits that several sightings made by Gemini astronauts could not be adequately explained. Gordon Cooper, in a letter to the United Nations, encouraged their involvement in the investigation of UFOs: "There are several of us who do believe in UFOs and who have had occasion to see a UFO on the ground, or from an airplane" [18]. In a recent interview, astronaut Buzz Aldrin describes an unidentified object that followed Apollo 11 out to the Moon in 1969[5]. Reports of UFOs continued, with a peak occurring in the mid 1970's.

Perhaps the best evidence for UFOs is an unusual event captured by one of the cameras aboard the Space Shuttle Discovery, mission STS-48, in 1991. Interpreting photographic evidence has always been somewhat problematic given the possibility that the photographs could have been be altered. Often, much of the time and effort in an UFO investigation is spent determining whether or not the incident in question could be a hoax. But a video shot by a US spacecraft virtually eliminates the possibility that the data could have been fabricated.

The event in question occurred on September 15, 1991, near the west coast of Australia. As many as a dozen objects moving in different directions relative to the space shuttle were involved. At the start of the event, an object appears at a point near the horizon and moves in a path along the horizon for a considerable distance (Figure 11). After a flash, the object abruptly changes direction and speed. This is followed a few seconds later by a streak that moves rapidly across the field of view and crosses the path of the object. Toward the end of the event, the camera looks on several objects moving below the shuttle (Figure 12). Zooming in on one reveals that it has a triangular shape. According to NASA [19]:

> The objects seen are orbiter-generated debris illuminated by the sun. The flicker of light is the result of firing of the attitude thrusters on the orbiter, and the abrupt motions of the particles result from the impact of gas jets from the thrusters.

---

[5] http://www.youtube.com/watch?v=XlkV1ybBnHI

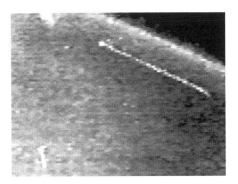 

Figure 11 Two time exposures constructed from a video taken during Space Shuttle mission STS-48 provide evidence of UFOs in space. One of the objects suddenly appears, and moves along the horizon line for a considerable distance (left). Shortly after a flash, it abruptly changes direction and accelerates rapidly (right).

Although this explanation seems reasonable at first, it does not hold up under scrutiny as Jack Kasher, a professor of physics and astronomy at the University of Nebraska in Omaha, and I have both shown. In Kasher's analysis, the movement of the objects was found to be inconsistent with the flow of gas away from the thruster, suggesting their sudden change in motion could not have been caused by a thruster firing.

One of the objects in the video appears at a point near the horizon, some 2700 km away. At this distance, prior to the flash, its speed was about 25.8 km/sec, which is four times faster than it would be moving if it were in orbit. After the flash, the object abruptly changes direction and accelerates within seconds to a speed of 400 km/sec — an acceleration of about 20,000 Gs!

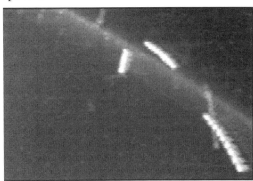 

Figure 12 Strongest evidence that some of the objects in the STS-48 video are not ice particles but UFOs. Curved paths to the right indicate the objects are far from shuttle and moving at considerable speed around the Earth (left). Close up of object photographed below the shuttle (right).

Perhaps the most compelling evidence that these objects are UFOs is their motion over time. Objects near the shuttle move in straight lines, unless they are acted upon by an external force such as the gases from a thruster firing. But a number of objects in the video move in paths that follow the curvature of the

Earth (Figure 12). This can mean only one thing: that they are far from shuttle and moving at a high rate of speed, estimated to be about 35 km/sec.

The tremendous speeds and accelerations implied in the video suggest the existence of aerospace technology well beyond the state of the art (as far as we know it). This is something that we have learned from studying UFOs, and only through further study will we learn more. Contrary to the Condon report, there is much to be learned from UFOs as this brief example and other reports published since 1968 have shown. Whether or not the findings of the Robertson panel, the Brookings Institute report, and the Condon study are relevant today, their recommendations are of concern and warrant further consideration.

The Robertson panel suggested various means be used to reduce interest in UFOs. Although it can be argued that, if anything, mass media has only served to increase public interest in UFOs, it can also be shown that serious treatment of the subject is lacking. Most mainstream scientific treatments of UFOs and related phenomena follow the formula laid out by the Robertson panel, using "actual case histories which had been puzzling at first but later explained." They usually have a skeptical expert, often an astronomer, giving the scientist's perspective.

Although the Condon study does not discourage UFO-related research, it does nothing to encourage it. The parallel channels of funding that may have once existed within the federal government to prevent errors in judgment made by any one school of thought simply do not exist today. Under fiscal pressure, government accountants do their best to eliminate wasteful overlap between different agencies doing the same thing. So when it comes to a field of study, say space science, NASA is the dominant funding agency. And within NASA, research in say, unmanned planetary exploration, is managed exclusively by Caltech's Jet Propulsion Laboratory. Furthermore, because of the close working relationship between government-funded scientists and their sponsors, the suppression of an alternative point of view does not have to be overt. Over time, censorship becomes institutionalized by sponsors who only fund certain types of research, by researchers who only study certain 'legitimate' subjects, and by journals who only publish results in well-established and agreed upon areas of investigation.

Perhaps the most troubling possibility suggested in the Brookings report is that the government might decide to withhold information regarding extraterrestrial contact from the public. Reading between the lines, it also suggests the possibility that scientists and decision makers might also be involved. We will not dwell on these possibilities further; however, one should keep them in mind over the course of the next few chapters, as the story of the discovery and investigation of possible extraterrestrial artifacts on the surface of Mars unfolds.

The Search for Life on Mars Resumes

As we prepared to land the first men on the Moon in the 1960's, scientists at the Jet Propulsion Laboratory in Pasadena California began exploring the solar system. In July 1965, Mariner 4 became the first spacecraft to photograph Mars, flying to within 6118 miles of the Red Planet. Even though only 19 photographs were obtained, the results were unexpected. Three of the pictures showed a cratered surface that looked more like the moon than Earth. Similar pictures were returned by Mariners 6 and 7, which passed within 2200 miles of the planet in the summer of 1969. As the spacecraft flew by Mars and its radio signals passed through its atmosphere, scientists discovered the atmosphere to be even thinner than expected.

After the launch failure of Mariner 8, Mariner 9 reached Mars on November 14, 1971 to find it blanketed by a global dust storm. Where earlier probes had flown past the Red Planet, Mariner 9 was the first to orbit it. When the dust finally cleared, a completely different picture was beginning to emerge from those of earlier fly-by missions. Photographs returned by Mariner 9 showed enormous shield volcanoes — the largest, Olympus Mons, is almost three times as tall as Hawaii's Mauna Loa, a vast canyon system, Valles Marineris, stretching almost one quarter of the way around the planet, and extensive networks of channels and tributaries, which strongly suggested that Mars, at one time in its history, had a great deal of liquid water flowing on its surface.

As the early Mariner probes transformed the late 19th century view of Mars as a planet crisscrossed with canals to one more like the Moon, a few years later Mariner 9 seemed to change it back again. In 1975, two Viking spacecraft were sent to Mars to follow-up on the discoveries of Mariner 9. Each consisted of an orbiter and lander. As the orbiters mapped the surface of the planet, the landers would attempt to soft land and perform a series of experiments to determine if microbial life existed in the Martian soil.

Although most planetary scientists are of the opinion that neither the orbiters nor the landers found any evidence of life on Mars, closer examination of the Viking data would suggest otherwise.

Sources

1. H.G. Wells, *The War of the Worlds*, Fawcett Publications, Greenwich, CT.
2. *Briefing Document: Operation Majestic 12*, Prepared for President-Elect Dwight D. Eisenhower, 18 November 1952 (http://www.majesticdocuments.com).
3. Stanton T. Friedman, *Top Secret/Majic*, Marlowe and Company, New York, 1996.
4. "Report of Meetings of Scientific Advisory Panel on Unidentified Flying Objects" (Robertson Panel) 14-18 January 1953, in *Scientific Study of*

Unidentified Flying Objects by Edward U. Condon, Bantam Books, New York, 1968.

5. Willy Ley, *Rockets, Missiles and Space Travel*, Viking Press, New York, 1951.
6. C. D. Jackson and R. E. Hohmann, "An Historic Report On Life In Space: Tesla, Marconi, Todd," *American Rocket Society 17th Annual Meeting*, Los Angeles, CA, November 13-18, 1962.
7. James W. Deardorff, "Possible extraterrestrial strategy for Earth," *Quarterly Journal of the Royal Astronomical Society*, Vol. 27, pp 94-101, 1986.
8. Carl Sagan, "Direct Contact Among Galactic Civilizations by Relativistic Interstellar Flight," *Planetary and Space Science*, Vol. 11, pp 485-498, 1963.
9. "Mysterious statuesque shadows photographed on Moon by Orbiter, *Washington Post*, November 23, 1966.
10. "Regular geometric patterns formed by Moon 'spires'," *Boeing News*, March 30, 1967.
11. George Leonard, *Somebody Else is on the Moon*, Pocket Books, New York, 1977.
12. Fred Steckling, *We Discovered Alien Bases on the Moon*," GAF International Publishers, Vista, CA, 1981.
13. Carl Sagan, "The man in the moon," *Parade Magazine*, June 2, 1985.
14. V. Arkhipov, "Earth-Moon system as a collector of alien artifacts," *Journal of the British Interplanetary Society*, Vol. 51, pp 181-184, 1998.
15. *Apollo Over the Moon: A View From Orbit* (NASA SP-362), Scientific and Technical Information Office 1978, NASA.
16. *Proposed Studies on the Implications of Peaceful Space Activities for Human Affairs*, Report No. 242, U.S. Government Printing Office, Washington DC, April 18, 1961.
17. Edward U. Condon, *Scientific Study of Unidentified Flying Objects*, Bantam Books, New York, 1968.
18. Letter to UN Ambassador Griffith from L. Gordon Cooper dated November 9, 1978.
19. Mark J. Carlotto, "Digital Video Analysis Of Anomalous Space Objects, *Journal of Scientific Exploration*, Vol. 9, No. 1, pp 45-63, 1995.

Four — Contact

> What you are now looking at is the first evidence of intelligent life beyond the Earth. — From Arthur C. Clarke's, *2001: A Space Odyssey*

Like H.G. Well's, *The War of the Worlds*, Arthur C. Clarke's epic *2001: A Space Odyssey* is about what actual contact with an extraterrestrial intelligence might be like. A magnetic anomaly is discovered on the moon, near the crater Tycho. Shrouded in secrecy, it turns out to be a monolith, the same monolith that has mysteriously guided our evolution from primitive ape to modern man.

Ten years after *2001* was published, a Viking orbiter spacecraft circles Mars. It is late July, 1976. The first of two Viking landers just touched down on the Martian surface a few days earlier. Now its orbiting mother ship searches for a suitable site for a second lander, due to arrive in a little over a week. Pictures are pouring in from Mars and are being displayed on video monitors at JPL in real time as they are received.

It too is summertime on Mars. On the thirty-fifth orbit, the Viking camera takes a picture over a region on Mars known as Cydonia. Cydonia is located in a transitional zone between cratered highlands to the south, and smooth low-lying plains to the north. It is late afternoon. At first glance, the picture seems to contain the usual assortment of mesas and other landforms typical for this part of Mars. But then, Toby Owen, a member of the Viking imaging team, notices something very unusual near the middle of the photograph, something that looks like a humanoid face staring up from the surface of Mars, and remarks, "O my God, look at this!"

Terrestrial Mind Set

A few years earlier, the possibility of finding advanced forms of life on Mars had actually been considered by none other than Carl Sagan. In 1971, Sagan and colleague David Wallace published a paper on the results of a search for life on Earth using satellite imagery [2]. After studying several thousand photographs taken by Gemini and Apollo astronauts, they derived visual criteria for detecting signs of life at different stages in its development. Sagan and Wallace reasoned that to detect non-intelligent life (e.g., life on Earth millions of years ago), a fairly extensive search of a planet's surface using high-resolution imagery (up to 1 meter per pixel) would be required. On the other hand, the presence of a highly advanced technical civilization would be relatively easy to detect, assuming the inhabitants had extensively modified the surface of their planet. Looking for the familiar signs of contemporary human habitation — the rectangular patterns of agricultural fields and urban areas, roads, canals, jet contrails, and industrial pollution — would require something between the two.

Figure 13 NASA's original picture of the Face on Mars released in July 1976. (NASA/JPL)

They concluded that indications of a biology on Mars comparable to that on Earth today would not have been detected by previous probes, i.e., by Mariners 4, 6, and 7, but go on to say that: "Forthcoming Mars orbiter and lander imaging experiments [Mariner 9 and Viking] hold significant promise of detecting life on Mars of contemporary terrestrial extent and advancement, should such life exist."

But one possibility Sagan and Wallace did not considered in their study was this: what if Mars was once inhabited, and has on its surface structures that were abandoned long ago, structures that have undergone extensive erosion, some so eroded that they almost seem to blend into the landscape. Would such artifacts be detectable, and more importantly, would they be recognized for what they were?

No one in 1976 expected to find archaeological ruins, let alone a humanoid face, on Mars. Most believed that Mars was more like the Moon than Earth, that it has been a dead planet for billions of years, dead long before the human race even began. So, no one thought it was unusual when, later that day, Viking project scientist Gerry Soffen showed a picture of the 'Face' to the press (Figure 13) and said "Isn't it peculiar what tricks of lighting and shadows can do. When we took the picture a few hours later, it all went away," or, when NASA released a photograph of the Face along with this brief statement a few days later:

> This picture is one of many taken in the northern latitudes of Mars by the Viking 1 Orbiter in search of a landing site for Viking 2. The picture shows eroded mesa-like landforms. The huge rock formation in the center, which resembles a human head, is formed by shadows giving the illusion of eyes, nose and mouth...

Richard Hoagland, then a member of the press corps, was there. He remembered it this way: "Gerry Soffen was a very open, very careful, engaging project scientist who typified the spirit around Viking, which was a multidisciplinary, open, American approach to probing the unknown... So when he said that nothing was there — that it was a trick of light and shadow — his credibility was overwhelming and certainly dissuaded anyone from doing any hindsight checking. We believed him" [3].

Rediscovery

After the Face was discovered in the summer of 1976, it appeared from time to time in the popular press. In 1977, while looking through a magazine, Vincent DiPietro saw a picture of the Face. The caption said that it was taken by a Viking spacecraft the previous year. Thinking the article was a joke, he put it away and forgot about it.

Two years later, DiPietro came across the same image, this time in the photographic archives of the National Space Science Data Center (NSSDC). NSSDC located at the Goddard Space Flight Center, just outside Washington DC, is the repository of the nation's planetary science data.

> There before me in black and white was the same serene image of a human-like face against the background of the Martian land surface. The title was certainly not misleading; it simply said 'HEAD'... At this point, I knew the object was not a hoax or it would not have been so boldly displayed in the NASA archives. I felt relieved and inquisitive; relieved that NASA had noted the picture and would presumably have verified it, and inquisitive to want to know more. But there was nothing more [4].

Realizing no one else at NASA had taken notice of this image, designated 35A72, DiPietro and his colleague, Gregory Molenaar, worked to improve the image quality. The Viking photographs are identified by their image ID: 35A72 is the 72nd picture by the 'A' spacecraft (i.e., Viking 1) during its 35th orbit. Since the Face occupied only a small portion of the image, DiPietro and Molenaar applied a technique they had developed to enhance and enlarge Earth satellite images. Their technique, known as the Starburst Pixel Interleaving Technique or SPIT, increases the spatial resolution of an image (i.e., the number of picture elements or pixels) by sub-dividing each pixel into a nine sub-pixels arranged in a 3 by 3 square. SPIT improved the visual quality of the data and seemed to enhance the impression of a face. But they needed another picture to verify that what they saw was real.

Soffen said another picture was taken several hours later that showed only an ordinary mesa. He told the press the Face had disappeared. How Soffen could have stated the orbiter was again over Cydonia a few hours later is, in itself, a mystery. Both of the Viking orbiters were placed in nearly synchronous, 24-hour orbits. Viking 1 was in an elliptical orbit inclined 39 degrees north of east for most of the mission. Its closest approach to Mars, which is known as periapsis, was at about 1500 km. Twelve hours later, the orbiter was at its farthest point from Mars (apoapsis), about 33,000 km away. When 35A72 was shot, the orbiter was about 1800 km above Mars. It had just passed periapsis, and was moving to the northeast, toward apoapsis. Several hours after 35A72 was snapped, Viking was thousands of kilometers away from Cydonia.

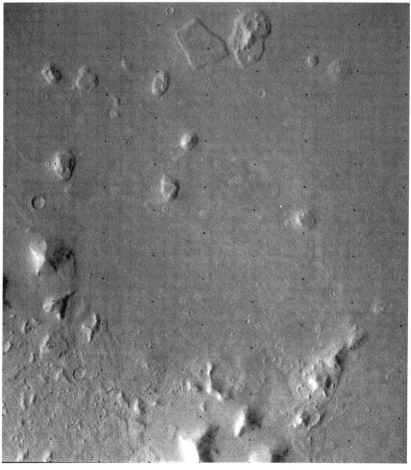

Figure 14 Second image, 70A13, containing the Face (upper left) and D&M Pyramid (left center) found by DiPietro and Molenaar. (NASA/JPL)

So it was no surprise when DiPietro and Molenaar looked through the archives for pictures taken a few hours later, they found nothing. But they kept looking, and eventually did find a second image, 70A13, taken not a few hours, but 35 days later (Figure 14). As Mars moved in its orbit around the sun, the position of the sun at periapsis changed about 1/2 degree per day. With this second picture in hand, there was no way the Face could be a trick of light and shadow as Soffen had said. Not only did it confirm the general features seen in the first image, but because the picture was taken 35 days later with the sun about 17 degrees higher in the sky, it revealed more of the right side of the Face. Showing evidence of an eye in what was shadow in the earlier image, this second image revealed the Face to be a reasonably symmetrical object.

Extending their search for similar features in neighboring frames, they found none. But to the south, DiPietro and Molenaar discovered an enormous pyramidal object. Noting that Mariner 9 had imaged pyramid-like objects on another part of Mars, they state "Of all observations of pyramids on Mars, we find that this one is the most unusual... and it is within ten miles of the Face." They go on to describe the pyramid as being five-sided, but say that it is difficult to tell because the confirming details are obscured.

They presented their results at the *Annual Convention of the American Astronomical Society* in College Park, Maryland, in June, 1980. Almost one thousand people, mostly scientists, attended the meeting. Although the response to DiPietro and Molenaar's talk was mostly positive, several criticisms were made. One noted that face-like rock formations do occur in nature, citing one in South America. Others had questions about the surrounding landforms. For example, could the pyramidal objects have been formed by tectonic processes, or by wind erosion?

A City on Mars?

Hoagland met DiPietro and Molenaar the following year at the *Case for Mars* conference in Boulder, Colorado. Hoagland too became interested in the Face,

> ... I realized that I was looking at something that was either a complete waste of time, or the most important discovery of the twentieth century if not of our entire existence on Earth [3].

Hoagland began to see that what DiPietro and Molenaar had discovered was but a piece of a larger puzzle. In particular, his attention was drawn to a collection of pyramidal objects to the southwest which he dubbed the 'City' (Figure 15). These features reminded him of Paolo Soleri's archologies — an idea for housing large urban populations in three-dimensional pyramidal structures.

> My original work on the location of the City came as a result of applying some basic cultural questions to the problem posed by the existence on the Martian surface of the Face. If this is, indeed, a monument of monumental proportions, then someone

had to have a reasonable length of time to carve it, sculpt it, or construct it... This raised in my mind the obvious question: where did all those people live?

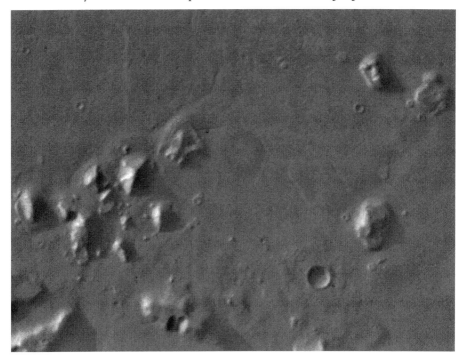

Figure 15 Digitally restored and contrast-enhanced subscene from 35A72. The image is about 33 by 27 km in area and is oriented so that north is up. This view shows the Face along with a collection of polyhedral objects to the southwest known as the 'City'.

Hoagland described the City as a "collection of objects located about 8 miles southwest of the Face, of which there are at least two truly suspicious and very artificial looking representatives: the 'Fort' [the first object southwest of the Face], and the 'Starfish Pyramid' located immediately southwest of the Fort. Other strange objects belonging to this City include a collection of five small objects located in the heart of the complex, a set of small domes and cones scattered non-randomly through this small area, a definite rectilinear arrangement of small pyramidal objects west of the starfish, and a long bright structure arranged at right angles to the southeast wall of the starfish-shaped object."

Concerning the object, which he called the 'Fort'

> Its strangeness was highlighted by two features apparently not shared by any other object in the photograph: two walls that met at almost a right angle at the southeast corner; and an inexplicable keep — a decidedly square-looking dark space — contained by these outer walls.

In addition to the above features, which are clearly resolved in the imagery, he detected a subtle pattern resembling a fine mesh to the south of the Fort, which he called the 'Honeycomb'. The Honeycomb would later be called into question as an imaging artifact.

Solstice Hypothesis

Hoagland noticed the City appeared to be aligned with the Face — in particular, that a line through the mouth points at the City. He reasoned that the City would be an ideal location to view the Face in profile. Another good location was from an odd-looking landform on the opposite side, the 'Cliff', which is next to an impact crater northeast of the Face. He measured the angle of a line from the mouth of the Face to the City. Its value, 23.5 degrees north of east, suggested something interesting: the possibility that the alignment might be solsticial, possibly providing a clue as to how old this collection of objects might be.

Sir William Herschel first measured the axial tilt of Mars in 1783, and suggested that Mars, like Earth, had seasons. The point on the horizon where the sun rises and sets changes with the season. On the first day of autumn (fall equinox), the sun rises due east and sets due west, regardless of latitude. With each passing day, the sun rises and sets farther to the south, making increasingly shorter sweeps of the sky. The days get shorter and the nights longer. On the first day of winter (winter solstice), the sun rises at its maximum point south of east and sets at its maximum point south of west. The cycle continues. Day by day, the sun rises less south of east until on the first day of spring (spring equinox), it again rises due east. With each passing day, the sun rises and sets further to the north making increasingly longer sweeps of the sky. The days get longer and the nights shorter. On the first day of summer (summer solstice), the sun rises at its maximum point north of east.

The angle marking the summer solstice sunrise depends on the latitude of the observer and the axial tilt of the planet. Today the tilt of Mars' axis is about 25 degrees. Hoagland calculated that the location of the summer solstice sunrise today on Mars, at the latitude of the City and Face (about 41 degrees N), is 34.9 degrees north of east. This is a point considerably farther to the north than the angle of the alignment he had measured from the Viking imagery. Because the tilt of Mars' axis changes, at some point in time it would have risen 23.5 degrees north of east. Hoagland determined that the last time this occurred was about 500,000 years ago. If the City and Face were artificially-constructed objects, he concluded they must be at least half a million years old.

The Independent Mars Investigation

Curiously enough, no one other than Hoagland had shown much interest in DiPietro and Molenaar's work. In the summer of 1983, anthropologist Randy Pozos, Ren Breck, vice-president of a computer conferencing company, and Richard Hoagland decided to organize a computer conference (similar to today's newsgroup) to find out what other scientists thought about the Face and the idea of a lost civilization on Mars. Hoagland and Pozos had different, but complementary, goals for the conference. For Pozos, the goals were somewhat general and open-ended [5]:

> The Mars project provides several major opportunities to advance research in the central human questions which underlie the humanities and the social sciences.
>
> Apart from the question of whether the landforms are the product of natural or intelligent process, the question of other intelligent life forms in the universe raises questions about the nature of consciousness...

Hoagland, on the other hand, had very specific goals in mind. In his opening remarks, he states:

> This conference is designed to link a hand-picked group of individuals, covering the widest range of scientific and social disciplines, in an effort to determine if this discovery [that of a City and Face on Mars] is, indeed, valid. Its secondary purpose, if the validity is established, will be to discuss the implications inherent in this discovery and to suggest profitable directions for further investigation.

Hoagland strongly believed the objects to be artificial and that a follow-up probe should be sent to Mars as quickly as possible to find out:

> There is a bigger picture here. Our purpose should not be to prove there was once life on Mars, merely to make so strong the circumstantial case that something very strange happed there that we go back — with the technology which can answer once and for all these fascinating questions. And we have less than two years — until the Martian window opens. My objective is to have built a 'Grand Jury case' that will permit the sending of the Galileo II hardware [an idea Hoagland had had for a quick return to Mars] back to an orbit of the planet, for such detail is required to resolve these pyramiding coincidences.

The computer conference began in December 1983. The group initially consisted of DiPietro and Molenaar, Lambert Dolphin, a physicist from SRI who had done extensive research work on pyramids and lost cities in Egypt and the Middle East, Gene Cordell, a computer and imaging specialist, Hoagland, and Pozos. A few months later another physicist, John Brandenburg from Sandia Labs, joined the conference.

For the most part the discussions fell into four general areas: the evidence for and against artificiality, research methodologies (ways of approaching the problem of determining whether or not objects on another planet might be artificial), interpreting the evidence, and implications of the potential discovery of life on Mars.

Evidence of Artificiality

Hoagland initially placed a great deal of emphasis on a very subtle honeycomb-like pattern he noticed in the City. However, Cordell's analysis of the pattern suggested that it probably was an imaging artifact. Hoagland also believed the alignment of the Face and City to be a key piece of evidence — "that there was a planned astronomical alignment inherent in their planning, design, and execution on the Martian surface." Dolphin saw it another way:

> The Hoagland hypothesis is surely a possibility, but it sounds a bit 'von Daniken' to me until we establish the City's existence beyond doubt. I agree that the alignments with the polar axis, sunrise, etc. is good evidence to include and that archaeoastronomy on Earth is a valuable new science.

In other words, using the solsticial alignment as evidence was premature until the artificiality of the objects involved in the alignment, namely the City and Face could be established by other means.

In the meantime, John Brandenburg found other images of the Face and City. They were, unfortunately, much lower in resolution than 35A72 and 70A13. However, one of them, 753A33, was acquired in the morning with the sun illuminating the Face from the right. Recall that 35A72 and 70A13 were both taken in the afternoon with the right side of the face in shadow. Although the resolution of 753A33 was not sufficient to make out any facial detail (Brandenburg estimated the eye socket to be on the order of a pixel in size), his impression was "that the photo confirmed the overall symmetry of the head, including its supporting structure. The Face was also shown to be framed completely around..."

The opinion that the Face was not only symmetrical, but also well-proportioned and precisely executed was expressed by James Channon, an artist in the group. Concerning its proportions he states: "The artist uses classical proportions and relationships when constructing the human face... The physical anthropologist recognizes a set of classic proportions, that relate facial features in predictable ways. The features on this Face on Mars fall within conventions established by these two disciplines." Channon was even more impressed however by the expression of the Face:

> For the artist, there is yet a more precise way to judge the authenticity of this form. The expression expected from one powerful enough to be so memorialized by a monument of this scale would not be random. The artistic, cultural, mythic, and spiritual considerations behind such a work of art would demand a predictable expression. The expression of The Face on Mars reflects permanence, presence, strength, and similar characteristics in this range of reverence and respect.

Other than the Face, the only object to receive much attention during this part of the investigation was the large pyramid discovered by DiPietro and Molenaar south of the Face. Hoagland's analysis of the D&M Pyramid was quite detailed.

He saw it as a five-sided structure, which was carved, not built, from a pre-existing landform. Dolphin, on the other hand, saw it as four-sided with one badly damaged face. Given the lesser gravity on Mars, Dolphin also thought that, in principle, such a structure could have been built. Hoagland estimated the height of the object to be approximately one mile above the surrounding terrain, with sides sloped at about a 30-degree angle. Both noticed that one of the sides appeared to be damaged.

> The substantial evidence that the southeastern side (matching the 1.6 mile northwestern side) has collapsed, with its material flowing out from the base in a fashion [is] very similar to the famed Egyptian pyramid at Meidum. This catastrophic event was triggered, I believe, by whatever made the peculiar small crater near the northeast 'buttress' still visible above the debris...

Research Methodology

Because of who we are, we interpret reality in terms of human values and experience. It is inescapable. We regard mankind as the center of the universe. According to Pozos:

> Anthropologically, the proper analysis of these Martian landforms challenges our ability to reach beyond the conceptual limitations of our species. Generally, our criteria for evaluating — let alone the perceptual structuring — of these landforms and other data sets is based on anthropomorphic criteria developed on one planet.

For Pozos, the fundamental question was how do we develop and test hypotheses about extraterrestrial artifacts so as not to be biased by our anthropocentric and geocentric experiences.

> In arriving at scientific truth there are generally two approaches. The first is to propose a hypothesis and then look for supporting evidence. The second is to amass evidence and look for patterns or trends from which to deduce laws or probable causes.

The scientific method employs two kinds of reasoning: deductive reasoning, which is the process of generalizing from a large number of facts, and inductive reasoning, which moves from a limited number of facts to broad generalizations. Hoagland reasoned inductively that if the Face is artificial, someone had to build it, and the builders had to live somewhere, probably some place nearby. The City provided a justification for the Face. That the City and Face could be artificial was further supported by an alignment, which seemed to suggest the objects were built at a time when Mars was more habitable than it is today.

Dolphin felt Hoagland's hypotheses were premature:

> .. I feel other examples of interesting landforms on Mars should be gathered and studied prior to formulation of theories about their apparent origin. The more

evidence gathered the better the case. If formulation of explanations is delayed, one is usually less likely to be biased by a special theory explaining the features and their associated intelligent life-forms...

The group expanded their search for unusual landforms beyond the City and Face, both within the immediate area, as well as to other parts of Mars. The continuation of a line from the City through the Face led to an unusual object located next to an impact crater known as the 'Cliff'. Hoagland conjectured that the Cliff could have served as a backdrop for the viewing of the Face from the City. Although this idea was later dropped, the Cliff remained an anomaly due to its location next to an impact crater that lacked any debris flow over or around it — suggesting the possibility that the formation or construction of the Cliff post-dated the impact. Hoagland and Brandenburg also noticed other unusual features nearby, including a low relief feature, later called the 'Tholus'.

Hoagland observed that the City and Face are located near the '0 km datum' — what would, in effect, be sea level if Mars had water. He suggested a search for other objects be conducted along this hypothetical Martian shoreline. With that as a working hypothesis, several other unusual objects were found to the southwest including a large pyramidal landform oriented almost north-south in frame 219S16, and several very small objects ('Gate Pyramid' and 'Pentagon') in 72A14. Being isolated features and not particularly unusual in comparison with the City and Face, they received little attention.

The group also found an unusual pyramidal object located on the rim of a crater in the Deuteronilus region to the northeast (Viking frames 43A01-04). The object, oriented 45 degrees relative to the compass directions, is located almost half way between the equator and the North Pole at 46.3 N, 353 W and is the highest point for more than 100 km in all directions. They also noticed a pattern of parallel grooves in the ejecta blanket of a nearby pedestal crater.

Perhaps the most interesting feature outside of Cydonia was found on the other side of the planet, on the slopes of Hecates Tholus (frame 86A08). Termed the 'Runway', the feature consists of a series of bumps, each about 300 meters tall, spaced 300 meters apart in a line about 4 km long.

Preliminary Conclusions

Based on the evidence reviewed, Dolphin believed the probability that the objects are artificial to be about 25-30%. Hoagland was more optimistic and saw it closer to 80%. Bill Beatty, a senior geologist who worked with Dolphin at SRI estimated the probability at 5%. According to Dolphin:

> Geologist Bill Beatty notes that the face is similar to other 'bulges' in the Cydonia area and if one were looking for a natural explanation, he suggests impact craters, slumping and 'capricious' wind erosion and deposition. If the face is not natural, then it is evidently carved from an existing hill as opposed to being built up out of blocks. He notes that wind erosion and deposition has been a major factor on the

Martian surface. If natural, the pyramids could be remnants or segments of volcanic craters occurring along near-vertical faults in the crust. The magma pushed up through vertical faults tends to 'blockiness', hence angular features.

Like Hoagland, Brandenburg also saw the Face as an artificial object, citing the following evidence:

> 1. It appears to be completely bisymmetric.
>
> 2. It has two eyes, a noise, and a mouth.
>
> 3. It appears to have an eye[ball] in one socket and also by my careful study, to have cheek ornaments below the eyes.
>
> 4. It is pleasing aesthetically, it looks like a king.
>
> 5. Other objects of non-natural appearance are found in the immediate area and elsewhere [on Mars].
>
> 6. This site and others appear to be areas where water was once abundant.

He also noted that "all objects resembling this object, found on Earth are man-made. I have never seen any natural formation anywhere or heard of any that appeared like this, resembling a face to these degrees."

During the course of the investigation, evidence that Mars probably once had a much thicker atmosphere, perhaps rich in oxygen, and possibly a vast northern ocean were discussed. However it was, and still is, widely believed that Mars lost its atmosphere early — during the first billion years, and that liquid water soon disappeared as well.

That a billion years would probably not be enough time for a technological civilization to evolve on Mars led Hoagland to the conclusion that if these structures are artificial they were probably built by visitors to Mars, either from Earth during a previous technological civilization, or from outside the solar system.

But Brandenburg saw another possibility:

> I myself supported another line of inquiry. I wanted to know if, given the past climate of Mars, indigenous intelligent life could have been possible there. Was there ever a time — a long enough time — when this could have happened?

Later on, we shall discuss current theories about the existence of life on Mars — past, present, and future — and examine both of these possibilities in greater depth.

Planetary Scientists Respond

After the computer conference, things went rather badly for the independent Mars investigation. The group submitted a paper, "The Preliminary Findings of the Independent Mars Investigation Team: New Thoughts on Unusual Surface Features" to the *Case for Mars II* conference held in the summer of 1984. It was accepted, but only as a poster paper, as a short ten-minute presentation off to the side. Hoping to engage members of the planetary community, their response was, according to Hoagland, "less than overwhelming." When Hoagland asked Chris McKay, one of the organizers of the conference, about joining in the investigation, McKay responded: "What could you possibly do with these images that NASA hasn't already done?"

The following July at the *Steps to Mars* conference held in Washington DC, journalist Jeff Greenwald had the opportunity to interview several planetary scientists to ask their opinion about the Mars anomalies.

Reactions ranged from cautious and diplomatic, to disinterested and outspoken. Geologist Harold Masursky said that among the thousands of mesas in Cydonia he would have been surprised if none looked like a face. He went on to say that "It's not that I don't take the investigation seriously, I just think there are more interesting features from a geologic standpoint" [6].

When asked about the report of the Independent Mars Investigation Team, Gerry Soffen, the Viking project scientist who stated the Face was a trick of shadow told Greenwald, "Most of us took fairly elementary looks at it." He went on to say, "I really haven't been that interested; and I'm still not."

Carl Sagan was the most opinionated:

> What they [Independent Mars Investigation Team] are proposing is to use existing — and some novel — computer enhancement techniques on existing data. Now, this area, like all other areas of interest on Mars, has already been subject to state-of-the-art enhancement... And there's nothing that comes out beyond what you've already seen.

When Greenwald asked Sagan if he thought the Face and City were worth investigation he stated:

> I'm not opposed to investigating. My view on the Face on Mars is my view on astrology. If someone can show that there is some validity to the claims, that's useful. But since the vast preponderance of the evidence is that it's nonsense, I don't think it's a good investment of resources.

Sagan and Hoagland were at opposite ends of the spectrum of opinion on Cydonia. For Hoagland, even the slightest chance these objects could be artificial was sufficient to justify an investigation. Sagan, like other planetary scientists, demanded extraordinary evidence. Perhaps the memory of Lowell's canals stuck in their minds. Perhaps the Face was just another figment of the human imagination.

Sources

1. Arthur C. Clarke, *2001: A Space Odyssey*, Signet Books, New York, 1968.
2. C. Sagan and D. Wallace, "A search for life on Earth at 100 meter resolution," *Icarus*, Vol. 15, pp 515-554, 1971.
3. Richard Hoagland, *The Monuments of Mars: A City on the Edge of Forever*, Frog Limited/North Atlantic Books, Berkeley CA, 1996.
4. Vincent DiPietro, Gregory Molenaar, and John Brandenburg, *Unusual Martian Surface Features*, 4th Edition, Mars Research, Glenn Dale MD, 1988.
5. Randolfo Pozos, *"The Face on Mars: Evidence for a Lost Civilization?,"* Chicago Review Press, Chicago IL, 1986.
6. Jeff Greenwald, "Of Mars and Men," in *Planetary Mysteries*, Richard Grossinger (ed.), North Atlantic Books, Berkeley CA, 1986.

Five — The Evidence

> I can't pretend to assert anything as positively true (for that would be madness) but only to advance a probable guess, the truth of which every one is at his own liberty to examine. If anyone therefore shall gravely tell me, that I have spent my time idly in a vain and fruitless enquiry after what be my own acknowledgement I can never come to be sure of; the answer is, that at this rate he would put down all Natural Philosophy as far as it concerns itself in searching into the Nature of things: In such noble and sublime Studies as these, 'tis a Glory to arrive at Probability, and the search itself rewards the pains. — Christiaan Huygens

In the fall of 1984, *Discovery* magazine published a summary of the *Case for Mars II* conference in its September issue. There was no mention of Hoagland, Brandenburg, or the independent Mars investigation. Yet in the same issue, Carl Sagan discussed a joint U.S./Russian mission to Mars for various reasons, including the investigation of "enigmatic surface markings and regularly arrayed pyramids on a high plateau," which he goes on to say is "hardly evidence for some ancient civilization on Mars, but nevertheless worth looking into."

The following June, Sagan wrote a piece entitled "The Man in the Moon" for *Parade* magazine highly critical of those investigating the Face on Mars. He begins by stating that because we see so many faces in our lifetime, our brains have become extremely effective in extracting faces from the clutter of other details. As a result, we sometimes see faces where there are none, as in the 'Man in the Moon'.

> The experience provides fair warning that, for a complex terrain sculpted by unfamiliar processes, amateurs examining photographs at the very limit of resolution may be in trouble. Their hopes and fears, the excitement of possible discoveries of great import, may overwhelm the usual skeptical and cautious approach of science [1].

With this set up, he then shifts to the real point of the article — the Face on Mars.

> There is a place on Mars called Cydonia, where a great stone face a kilometer in size stares unblinkingly up at the sky. It is an unfriendly face but recognizably human. In some representations, it could have been sculpted by Praxiteles, the fourth century B.C. Athenian. It lies in a landscape where many low hills have been molded into odd forms, perhaps by some mixture of ancient mudflows and subsequent wind erosion. From the number of impact craters nearby, the face looks to be at least tens of millions and perhaps billions of years old.

Sagan cites a recent story in the tabloids about a Soviet scientist who claims to have found ruined temples on Mars, but does not mention the Independent

Mars Investigation's work, or their paper at the *Case for Mars II* conference. He does though make this rather oblique reference to Hoagland:

> An American science writer compares the Martian face to 'similar faces ... constructed in civilizations on Earth. The faces are looking up at the sky because they are looking up to God. Is it a remnant of an ancient, long-extinct human civilization on Mars? Might they have come to Earth and initiated life here? Could it have been constructed by alien visitors stopping on Mars for a brief interlude? Was it left for us? What does it imply about human evolution?

He concludes by saying that because it is easy to speculate about the Face, one must apply "only the most rigorous standards of evidence."

A Closer Look

I first learned about the Face on Mars in a February 1985 article in the *Boston Globe*. The article, "Was There Life on Mars?" was short but provocative:

> What appears to be the face of a monkey stares into space from the surface of Mars in a photograph taken in 1976 by U.S. Viking spacecraft. Scientists of the Mars Investigation Group at the University of California in Berkeley say the face and what appear to be the remains of four huge pyramids suggest the existence of an ancient civilization on the planet. Officials of the National Aeronautics and Space Administration, however, contend figures were formed by natural elements as wind and blowing sand. The face, incidentally, is about a mile long and three-quarters of a mile wide and is about six miles from the pyramid-like features.

I was puzzled. How could these objects be seen by one group of scientists as the remains of an ancient civilization, and by another as natural geologic formations? Curious, I called Berkeley and tracked down the Mars Investigation Group, not to space sciences or physics, but to the Center for Research in Management, and arranged a meeting with Tom Rautenberg, the Administrative Director of the project. As it turned out, Rautenberg would be in Boston in a few weeks, and would bring a set of computer tapes with imagery from JPL.

I was then at TASC, a privately-owned company north of Boston, specializing in government-sponsored research and development. Using techniques we had developed for processing Earth satellite images, I worked to digitally restore and enhance the Viking images [2]. The first step was to remove 'salt and pepper' noise — the random pattern of black and white dots in the imagery caused by data transmission errors. Sagan had referred to these as data 'drop outs' in his Parade article, "If we look more carefully at the image, we see a strategically placed 'nostril' is in fact a bit of lost data in the radio transmission from Mars to Earth." I wanted to be sure to remove all such artifacts from the imagery. Next, I used standard image enhancement techniques to increase the visual quality and contrast of the images. Because the features of interest were rather small, between 10 and 100 pixels in size, they had to be digitally enlarged.

The Face was obvious in the original photograph released by NASA in the summer of 1976; however, the photo contained very little detail because of the way it had been processed. The only features anyone appeared to take notice of were the data drop-outs, which had not been removed in the image. Relying on Sagan's article, one could easily dismiss the Face as a strangely-lit rock formation. However, it became evident to me that on closer examination there were many subtle details in the imagery that were not visible in the batch processed NASA photographs (Figure 16).

After I restored and enhanced the original images, several broad, regularly-spaced stripes across the Face could be seen. There was evidence of an eyeball in the left eye cavity, which DiPietro and Molenaar had detected. But there was more — Just at the resolution limit of the camera, there appeared to be a pair of thin crossed lines over the forehead that looked like a headdress or crown, and within the mouth was fine structure that looked liked like teeth. If the Face was a natural formation as claimed by NASA, these features should not be there. Moreover, the fact that they were present both in both 35A72 and 70A13 meant they were not noise or processing artifacts. They had to be real.

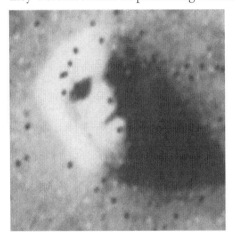 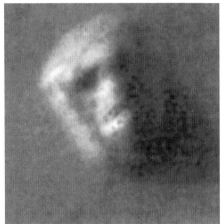

Figure 16 Close up of Face from the original JPL photo (left). In a more carefully restored image (right) internal details become evident.

A Trick of Light and Shadow?

In its 1976 press release, NASA stated that the Face "is formed by shadows giving the illusion of eyes, nose and mouth," in other words, an optical illusion. Although DiPietro and Molenaar did find a second image of the Face at a somewhat higher sun angle, it was still an afternoon image with the sun illuminating the Face from the left. With only these two images of the Face, NASA's statement was hard to dispute. But it occurred to me that if the shape of the Face could be recovered somehow, one could, in principle, generate synthetic images for different sun angles and viewpoints using computer graphics techniques. If it was an optical illusion as NASA claimed, these other computer-generated images should reveal an ordinary hill.

Purely by coincidence, in the summer of 1985 at TASC, we were developing computer vision techniques for producing height maps from imagery. Height maps (also known as digital elevation models, or DEMs) give the height or elevation of points on the ground and are used in a variety of commercial and military applications. Usually DEMs are produced from two images taken at slightly different positions in space. The shift in position of a feature (e.g., the top of a building) between the two images (parallax) is related to the height of the feature.

We were developing a different method known as shape-from-shading (or photoclinometry) to produce DEMs from a single image. Shape-from-shading exploits the relationship between the slope of the surface of an object and its reflectance, and hence, its brightness. Consider a smooth object in a dark room illuminated on one side by a flashlight. To an observer standing a short distance away, the side of the object facing the light is brightest; the opposite side is darkest, in shadow. Between the two sides there is a gradual decrease in brightness because the surface of the object is sloping away from the light and so a smaller and smaller amount of light is falling on the surface.

In computer graphics, 3-D models of objects are used to generate 2-D images. Shape-from-shading can be thought of as inverse computer graphics — determining the 3-D shape of an object from its 2-D image. Patrick Van Hove, a graduate student from MIT working at TASC that summer, implemented a shape-from-shading algorithm that worked well in recovering the shape of isolated objects. We tested the algorithm by taking a known object such as a hemisphere or pyramid, and generating a simulated image of it on a flat background. We then used shape-from-shading to determine the shape of the object from the image. By comparing the shape produced by the shape-from-shading algorithm to the original shape we were able to verify the accuracy of the algorithm.

In the meantime, along with Brian O'Leary, I began to apply shape-from-shading to the imagery of the Face. O'Leary, once a member of the astronaut corps, had written his Ph. D. thesis on the optical properties of Mars. Since there was no 'ground truth' to check the accuracy of our shape-from-shading results against, O'Leary suggested the surfaces computed from each image be used to predict what the other image should look. By means of this cross-check we were able to confirm that the computed surface was an accurate representation of the 3-D structure of the Face. Two key questions could then be addressed: Are the facial features seen in the imagery also present in the underlying surface, and do these features persist as the lighting conditions and viewpoint are changed?

The results were provocative: the computed 3-D structure clearly showed evidence of facial features in the mesa itself. Moreover, by generating views for other light source positions (other times of the day and year), and for other viewpoints (from the side as well as above), we could show that the appearance of a face persisted over a wide range of lighting and viewing positions (Figure 17). Unlike New Hampshire's Old Man of the Mountain and other facial profiles or silhouettes of natural origin, the Face was not an optical illusion as NASA had claimed.

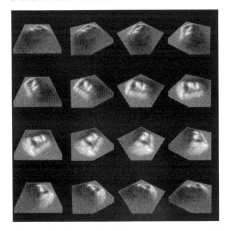 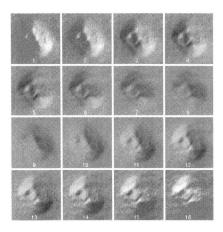

Figure 17 Once the 3-D surface of an object has been determined, synthetic views from any direction can be generated using computer graphics rendering techniques. In a series of perspective views from around the Face, features evident in the original overhead view can still be seen even when the object is viewed from other directions (left). Shaded views of the Face created by varying the light source direction (right). These simulated views looking down on the Face were for a summer day on Mars. Each frame is one hour apart.

Excited, O'Leary and I each wrote papers and submitted them to the planetary science journal *Icarus*. After several months, I received a letter from Joseph Burns, the editor of *Icarus*. Here is a part of his letter:

> I enclose two reviews on your recent submission, '3-D Digital Imagery Analysis of the Face on Mars'. Based on these reviews, I will not be able to accept the paper nor do I believe that revisions can make the paper acceptable. I agree with the arguments of the anonymous reviewer that the object is not of sufficient scientific interest.

One of the reviewers, Robert Wildey, who had done considerable work on photoclinometry, recommended the paper be accepted as a methodology paper. But the anonymous reviewer had this to say:

> The major point of this paper is that the hill in Cydonia, when viewed at low resolution, looks something like a face... The important scientific question is 'so what?' If he were foolhardy, the author could offer the hypothesis that the 'face' was carved by intelligent beings. This hypothesis is interesting, fanciful, and all but certain to result in rejection of the paper by any reputable scientific journal. It simply cannot be defended with the data in hand, and is not particularly plausible in light of everything we know about Mars. So, the author wisely sidesteps the issue, and states, quite correctly, that the resolution of the images is insufficient to shed further light

on the issue of the hill's origin. What are we left with then? We are left with a very detailed description of the topography of one among the millions of hills on Mars. Either the hill is interesting or it isn't. If the author had claimed the hill were interesting because it was an artifact of an ancient civilization, I would have recommended rejection on the grounds that it advanced a hypothesis for which it could not provide support. Instead the paper provides an unnecessarily detailed description of a feature that is not shown to be of any particular scientific interest. I do not recommend that the paper be published.

There was no way around his logic: based on everything he knew about Mars, there could not possibly be an artificial object there. In other words, because the Face on Mars cannot be there, it is not there. O'Leary's paper was also rejected on similar grounds. For O'Leary, who had published many papers before in *Icarus*, this was his first rejection.

Realizing any further dialog with the editor would be a waste of time, I resubmitted my paper to *Applied Optics*, a journal I had published in before. The paper was not only accepted, it made the cover of the May 15, 1988 issue [3]. Here are excerpts from two reviewers:

> This paper will undoubtedly be controversial because of its subject matter and conclusions, and it is probably not the usual fare for *Applied Optics*. I generally favor discussing controversial ideas in the literature rather than suppressing them, however, provided that they are well thought out and clearly articulated. For the most part, the author has been careful to explain and qualify his methods and results, and his conclusions and recommendations are modest and hard to quarrel with.
>
> I don't usually review manuscripts immediately upon receiving them, but this one 'grabbed me' after just a quick look... The methods of the investigation appear sound, the algorithms state of the art, and by an large the conclusions are consistent with the results of the study.

My findings were announced at a press conference organized by Richard Hoagland at the National Press Club in Washington DC a few months later. News stories that followed were skeptical but fair, saying that follow-up missions to Mars were needed to resolve the mystery.

Measuring Artificiality

Carl Sagan and I exchanged several letters during this time. After sending him an early draft of the shape-from-shading paper, he wrote back saying, "I certainly agree that Jerry Soffen's 'trick of lighting' off-hand remark is in error; that is, the impression of a face persists over a variation of all three photometric angles." In another letter, I mentioned that at TASC we were developing new techniques for detecting artificial objects in images based on fractal geometry, and that I was planning to apply these new techniques to the Viking imagery in order to assess the artificiality of the Face. He sent me an interesting paper he presented to the Royal Society in 1975 entitled "The Recognition of Extraterrestrial Intelligence." In it, he describes a way of detecting signs of extraterrestrial intelligence by measuring deviations from 'thermodynamic equilibrium'.

> Perhaps the most significant indication of intelligent life on Earth — and certainly the one most readily discernible — occurs in the radio part of the spectrum. Except for minor departures due to atmospheric absorption and emission, the electromagnetic emission of the Earth closely follows that of a black body... However in the radio part of the spectrum there is an immense departure from radiative equilibrium [4].

What Sagan had proposed was that deviations from natural background radiation could be used as an indicator of intelligence. The basic idea behind our technique of using fractals to detect artificial objects in imagery was very similar.

Generally speaking, nature tends to create structures that are 'self-similar' — structures in which a part resembles the whole in some sense. For example, over a range of scale, as one examines a leaf in greater and greater detail, the same kinds of patterns repeat at smaller and smaller scales. Objects that exhibit this kind of behavior are known as fractals. Fractal models can been used to describe a variety of natural phenomena besides plants, including clouds, lightning, and the shape of natural terrain and coastlines, to name just a few. Fractals have been used with great success to generate photorealistic terrain backgrounds for computer animations (recall the Genesis sequence from the movie *Star Trek II*). Artificial structures do not have this property, and so are not fractal.

Instead of using fractals to generate images of natural terrain backgrounds, our approach was the reverse — to try to model image backgrounds with fractals, the idea being that parts of images not modeled well by fractals could indicate the presence of an artificial object. A colleague of mine, Michael Stein, had developed a fractal technique for detecting manmade objects such as military vehicles in overhead imagery (Figure 18). Without any modification we applied the technique to Viking frame 35A72, and found that the Face was the least fractal object in the entire image!

Extending our analysis to other nearby frames, the Face remained the least fractal object over an area about 15,000 sq. km in size — more than five times the area of the state of Rhode Island. Results from Viking frame 70A13 corroborated this result. Several other objects within the City also were highly non-fractal.

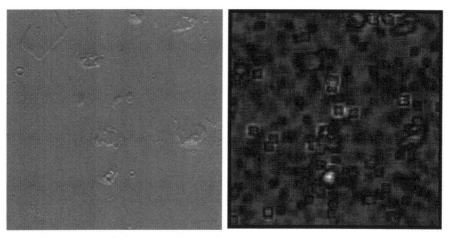

Figure 18 Fractal analysis results for the Face and surrounding area from 35A72 (left). Bright areas in the fractal model-fit image (right) indicate where image structure does not fit a fractal model and is least natural.

Figure 19 For comparison, an image over a U.S. military base (left) and its fractal model-fit error image (right). That the fractal technique detects artificial features and not simply differences in image texture is apparent.

Again, excited by these results, I decided to submit a paper, this time to the journal *Nature*. Although they were a little more polite than *Icarus*, they too rejected the paper for similar reasons. Persevering, our results were finally published in 1990 by the *Journal of the British Interplanetary Society* in a paper entitled, "A Method of Searching for Artificial Objects on Planetary Surfaces" [5]. O'Leary's earlier paper, which had been rejected by *Icarus*, also appeared in the same issue of *JBIS* [6].

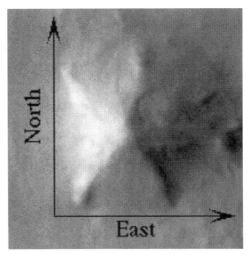

Figure 20 D&M Pyramid in frame 70A13 mapped to a Mercator projection. The south face of the pyramid appears to be aligned to the meridian.

Strange Geomorphology

Where my efforts were focused primarily on analyzing the Face, Erol Torun, a cartographer at the Defense Mapping Agency, was studying the D&M Pyramid — the pyramidal object DiPietro and Molenaar found south of the Face. Torun, who was trained in geology, systematically considered five natural explanations for the unusual faceted morphology of the D&M[6].

First, he considered fluvial deposition and erosion. According to Torun, "Fluvial processes can yield roughly symmetrical relief, such as the teardrop-shaped islands observed in many Martian stream beds. But in this case, fluvial processes can be ruled out as mechanisms for forming the D&M Pyramid as there are no indications that water ever flowed 1 km deep in Cydonia Mensae (1 km being the approximate height of the D&M Pyramid. It is also true that sharp edged multi-faceted symmetrical shapes are not characteristic of fluvial landforms."

In the popular TV series *Cosmos*, Carl Sagan called attention to a group of pyramidal objects photographed by Mariner 9 in 1972, which he called "The Pyramids of Elysium." According to Sagan, "The largest are 3 kilometers across at the base, and 1 kilometer high — much larger than the pyramids of Sumer, Egypt or Mexico on Earth. They seem eroded and ancient, and are, perhaps, only small mountains, sandblasted for ages. But they warrant, I think, a careful look." I mention these objects because they are sometimes confused with the D&M. The shape of the Elysium pyramids is characteristic of yardangs —

---

[6] http://users.starpower.net/etorun/pyramid/

landforms that have been modified by the abrasive action of wind-borne particulates. Yardangs typically have sharp edges aligned in the direction of the prevailing winds. Like Hoagland, Torun saw the D&M as a five-sided object. But whether it is four- or five-sided, the following assessment is valid:

> Five-sided symmetrical ventifacts or yardangs appear to be totally nonexistent on Earth and Mars. Prevailing winds are not likely to have shifted periodically with perfect symmetry and timing. Even if this seemingly impossible condition were satisfied, another factor would prevent such an object from forming. As noted above, locally reversed airflow can cut a flat surface perpendicular to the wind direction on the leeward side of a wind-cut hill. This locally reversed airflow, and associated surface level turbulence, would prevent the formation of this hypothetical five-sided ventifact. Each time the wind shifted to a new direction, the reversed airflow would start erasing the edges formed by other wind directions. The end result would not be a pyramidal hill, but rather a round one.

Next, Torun considered mass wasting — the downslope movement of large amounts of rock and/or soil under the influence of gravity — as a possible cause. Although he recognized the role of mass wasting in shaping many of the irregular knobs in Cydonia, he dismissed it as a mechanism for shaping the D&M:

> It is uncharacteristic of mass wasting of loose material, or slumping of single masses, for such material to slide off of a hill in such a way as to leave behind multiple flat surfaces and straight edges where none had previously existed... It is also unlikely that such mass wasting would occur symmetrically. When mass wasting produces a flat surface, it is normally due to some previously existing fault or sedimentation layer that provides a shear surface for the mass wasting or slumping. An analogous example from Earth geology would be the failure of a slope consisting of Cretaceous clay that has such internal layers. If this type of internal layering occurs on Mars, it is unlikely to occur symmetrically so as to yield a symmetrical erosional remnant.

That the D&M could be a volcano was eliminated as well. "There are no signs of significant volcanic activity in the Cydonia Mensae region, thus drastically reducing the possibility of any landform in the region being thus formed... Additionally, the D&M Pyramid has no vent at its apex, and exhibits a symmetry unknown in volcanic landforms."

He even considered the possibility that the D&M could be the result of some large scale crystal growth on Mars, but eliminated that too.

Tetrahedral Geometry

As time passed Hoagland, Torun, and later others become interested in certain mathematical relations that seemed to reoccur both within, and between selected objects in Cydonia. In a geometric reconstruction of the shape of the D&M Pyramid, Torun found that internal angles between the base and edges, and trigonometric functions of these angles could be expressed in terms of the mathematical constants π (the ratio of the circumference of a circle to its diameter, equal to 3.14159...), e (the base of natural logarithms, equal to 2.71828...), and the square roots of 2, 3, and 5. As described in the *McDaniel Report* [7], Kieth Morgan found Torun's model to be the "only pentagonal figure having two front angles of 60° that can represent the five constants $\sqrt{2}$, $\sqrt{3}$, $\sqrt{5}$, e, and π redundantly across angle ratios, radian measure, and trigonometric functions." It was also found that the ratio of the lengths of two of its internal dimensions was close to the Golden Ratio, which shows up repeatedly in growth patterns in nature has fascinated mathematicians and artists since the time of the Greeks. The Golden Ratio can be described as follows: Imagine a stick divided in such a way that the ratio of the longer part to the whole is the same as the ratio of the shorter part to the longer part. This ratio is the Golden Ratio, equal to $(1 + \sqrt{5})/2 \approx 1.61803$.

In a letter to Hoagland, it was pointed out that the latitude of the D&M Pyramid, about 40.868° N, is very close to the arc tangent of e/π, the ratio of two of the fundamental mathematical constants mentioned above [8]. The geomorphology of the D&M Pyramid was unusual, its reconstructed internal geometry elegant, and its latitude on Mars provocative. To Hoagland and Torun these recurring anomalies seemed more than just indicators of intelligent design, they suggested a 'message' — a message that was, in the spirit of the Pythagoreans and Gauss, mathematical.

The tetrahedron is one of five Platonic Solids (the others being the cube, octahedron, dodecahedron, and icosahedron) that consists of four triangular faces, four vertices, and six edges. Imagine placing a tetrahedron inside a sphere so that the four vertices touch the inside of the sphere. Torun determined the ratio of the surface area of the circumscribed sphere to that of the tetrahedron inside to be about 2.720669, which he called e'. That this value is close to e and the latitude line given by the arctangent of e'/π passes through the D&M Pyramid suggested the possibility that the message might have something to do with tetrahedral geometry.

If you turn a tetrahedron upside-down so that its base is up (north) and place it inside a sphere, the latitude of the three vertices containing the base is at about 19.5° N. The *McDaniel Report* provides additional evidence supporting Hoagland and Torun's claim that certain angles both in the D&M and between objects in Cydonia refer to this value. Motivated by these relations, Hoagland and Torun went on to develop a theory of 'hyperdimensional physics' — one that tries to explain the presence of 'emerging energy phenomena' on some of the planets and moons in our solar system — phenomena that exist near 19.5° N and S.

Because the significance of Hoagland and Torun's analysis is strongly dependent on the accuracy of the measurements, their results were wide open to criticism. Michael Malin, principal investigator for the Mars Orbital Camera (used in Mars Observer and Mars Global Surveyor) has been a particularly outspoken critic of the Cydonia investigation in general, and Hoagland and Torun's work in particular:

> Evidence cited as presently 'proving' these are unnatural landforms include measurements of angles and distances that define 'precise' mathematical relationships. One of the most popular is that 'The D&M Pyramid' is located at 40.868 degrees North Latitude...[7]

According to Malin, the latitude and longitude of the D&M can lie anywhere between 40.67 N, 9.62W and 40.71 N, 9.99 W, based on geodetic information derived from the Mars Digital Image Mosaic/U.S. Geological Survey control network and on Viking spacecraft tracking and engineering telemetry data. He goes on to say that:

> Even given accurate data, however, most science does not depend solely on planimetric measurements, even when using photographs. There are many other attributes used to examine features, especially those suspected of being artificial, and the Martian features do not display such attributes.

Later, we will learn about the kinds of things Malin would have to find on Mars to be convinced of their artificiality.

Testing For Random Geology

Torun's measurements were derived from a hypothetical model of the geometry of the D&M Pyramid reconstructed from the Viking imagery. Some features (edges and vertices) in his model do not correspond to well-defined features in the image. Of course this is to be expected if the object under consideration is degraded by wind and weathering. Similarly, many of Hoagland's measurements of various angles and relationship between the D&M and nearby objects such as the Face, Cliff, Tholus, and objects in the City were also based on poorly-defined features. That these measurements might not be accurate weakened the weight of the evidence for artificiality.

In addition to the larger features mentioned above, Hoagland noticed a number of smaller mound-like objects scattered throughout Cydonia. One group of three seemed to lie at the vertices of a right triangle. In 1994 Horace Crater, a professor of physics at the University of Tennessee Space Institute undertook a systematic analysis of 12 mounds including the three Hoagland had found. By focusing on the smaller mounds, which are distinguished from other objects in the scene by their uniformity in size, brightness, and shape, he hoped to reduce the arbitrariness of earlier analyses. Crater summarizes his approach as follows [9]:

---

[7] http://www.skepticfiles.org/skep2/marsobs1.htm

Investigation of the geometric relationships between these mounds takes the form of a test of what may be called the random geology hypothesis. This hypothesis presupposes that the distribution of the mounds... however orderly it may seem, is consistent with the action of random geological forces. Our question is: Does the random geology hypothesis succeed or fail in the case of the small mound configuration at Cydonia?

Instead of manually picking points to use as the basis of his measurements he developed a computer program to determine all possible triangles that could be drawn between 12 vertices. The actual positions of the vertices were allowed to vary within the areas of the mounds. The program systematically searched for all right triangles with angles 90°, 45°+t/2 and 45°-t/2, and isosceles triangles with angles 90°-t, 45°+t/2 and 45°+t/2 over all values of t from 0 to 90° in fixed increments. For each value of t, the program moved the vertices within the areas of the mounds and counted all triangles whose angles were within a given tolerance of the corresponding right and isosceles triangles. Crater first ran the program at a rather coarse 5-degree level of precision, and obtained a plot of the number of triangles found vs. t that was relatively flat. But when he tightened the precision of the test from 5° to 0.2°, the resultant plot had a strong peak of 19 triangles at 19.5° against an average background level of 5.7 triangles. He determined the statistical significance of this result to be extremely high — about 1 chance in 65,000 that the distribution of the mounds was random!

In their 1999 paper, "Mound Configurations on the Martian Cydonia Plain" published in the *Journal of Scientific Exploration*, Crater and co-author Stanley McDaniel concluded "that the random geology hypothesis fails by a very large margin, that a radical statistical anomaly exists in the distribution of mound formations in this area of Mars." Peter Sturrock, who was one of the referees said this in an editorial reply to their paper:

> Some geological formations are obviously very far from random... The authors are really testing whether the distribution is random. If they were to show convincingly that it is not, they may simply have shown that this array of mounds is another geological formation that is non-random.

Crater and McDaniel countered:

> The objection seems to assume that non-random geological formations are common, but this is not supported either qualitatively or quantitatively, nor is the method by which it would be determined stated. We are not aware of any studies, comparable to our own, carried out on terrestrial formations, that would identify such a degree of non-random distribution of separate objects. In discussions with geologists we have found none who could identify distributions of natural objects in terrestrial terrain displaying such non-random characteristics... Archaeologist J. F. Strange of the University of South Florida has stated ... that if the mounds had been found on the Earth, archaeological teams would be strongly motivated to investigate them. We therefore believe our results are substantive and indicate that further serious study of the region should be undertaken.

Reviewing the Evidence

The architectural design, facial proportions, and overall artistic impression suggested at the outset of the Independent Mars Investigation that the Face might be an artificial object. Subsequent tests of this hypothesis involving the enhancement of subtle detail in the Face, shape-from-shading/synthetic image generation to determine if the Face is an optical illusion, and fractal analysis to assess its shape in a quantitative manner have all provided cross-confirming evidence that support the original hypothesis. Other unusual objects have also been found nearby that appear to be related to one another.

In 1996, the Society for Scientific Exploration invited me to prepare a paper for their spring meeting at the University of Virginia in Charlottesville summarizing the evidence for artificiality in Cydonia. The paper which later appeared in the *Journal of Scientific Exploration* under the title: "Evidence in Support of the Hypothesis that Certain Objects on Mars are Artificial in Origin" was an attempt to examine all of the research results concerning the Face and nearby objects, and to determine if they were sufficient to support an extraterrestrial hypothesis [10].

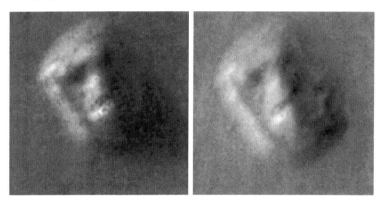

Figure 21 Two images of the Face from 35A72 (left) and 70A13 (right). The second image (70A13) found by DiPietro and Molenaar provides corroborating evidence for facial features, symmetry, and fine-scale detail.

Up to this point I had focused mainly on the Face. As I took stock of all of the evidence, I realized that it was substantial.

1) General Facial Characteristics — The Face on Mars possesses all of the features required of a humanoid face: head, eyes, ridge-like nose, and mouth. These features are present in two images taken at slightly different sun angles (35A72 and 70A13). In the original image, 35A72, the sun angle is only 10 degrees above the horizon and so most of the right side of the Face is in shadow. But in 70A13 the sun is 15 degrees higher, showing more of the Face's right side. Instead of an ordinary rock formation, this second image not only confirms the facial features first seen in 35A72, but also reveals the overall symmetry of the head, the extension of the mouth, and a matching eye socket on the right side.

2) Facial Proportions — As part of the independent Mars investigation, artist James Channon evaluated the Face in terms of its proportions, supporting structure, and expression. From both an artistic and anthropological standpoint he found "The features on this Face on Mars fall within conventions established by these two disciplines. I find no facial features that seem to violate classical conventions."

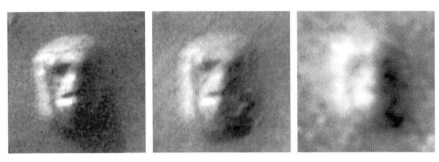

Figure 22 The three highest resolution Viking images of the Face from 35A72 (left), 70A13 (middle), and 561A25 (right) at 47.1, 43.3, and 162.7 meters/pixel, respectively.

3) Architectural Symmetry of Face Platform — Although the left and right sides do not match, the platform surrounding the Face is highly symmetrical. According to Channon, "The platform supporting the Face has its own set of classical proportions as well. Were the Face not present, we would still see four sets of parallel lines circumscribing four sloped areas of equal size. Having these four equally proportioned sides at right angles to each other creates a symmetric geometric rectangle." In addition to being impressed by its proportions and symmetry, Channon found the Face to have a strong aesthetic appeal emphasizing that "the artistic attention required to generate the expression like the one studied [i.e., the expression of the Face in 35A72] is not trivial."

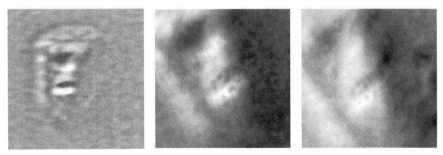

Figure 23 Image 35A72 enhanced to emphasize linear features (left). Note dark stripes or bands across the face and crossed lines above eyes. Fine structure in the mouth can be seen in 35A72 and 70A13 (middle and right). Since these features are present in two different images, they are not likely to be noise or processing artifacts.

4) Subtle Details of the Face — The Face contains a number of subtle details or embellishments including a dark cavity within the eye socket that looks like a pupil, broad stripes across the face, thin lines that intersect above the eyes, and fine structure in the mouth that look like teeth. These features are visible in both 35A72 and 70A13 and so cannot be noise in the imagery or processing artifacts. If the Face is the result of erosional processes as some claim, these same processes would also have to explain the subtle details — details that should have been obliterated by these processes over time.

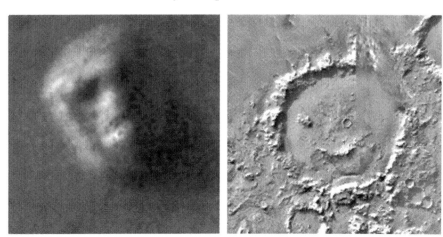

Figure 24 Comparison between the 'Happy Face' (right) and the Face on Mars (left) imaged by Viking. The 'Happy Face' has been used by NASA to illustrate the human tendency to see faces in nature.

5) Persistence of Facial Features — NASA maintains that the Face is "formed by shadows giving the illusion of eyes, nose and mouth," in other words, an optical illusion. Using shape-from-shading to determine the 3-D structure of the Face from the Viking imagery, and computer graphics techniques to predict how the computed 3-D structure appears under different lighting conditions and from other perspectives, we have shown this is, in fact, not the case. Shape-from-shading indicates that the facial features seen in the Viking image are also present in the underlying topography and produce the visual impression of a face over a wide range of illumination conditions and perspectives. These results have also been corroborated by the sculptress Kynthia who has created a model of the Face in clay that matches all available images under the corresponding light source/viewing conditions[8].

6) Fractal Analysis of Face — By using fractals to analyze terrestrial satellite images, manmade objects such as buildings and vehicles can be detected as statistical deviations from the natural background. Applying the same technique to the Viking imagery showed the Face to be the least fractal object in frame

---

[8] http://www.kynthia.net/marsF.html

35A72. When the analysis was applied to four surrounding Viking frames, it was the least fractal object over the entire area.

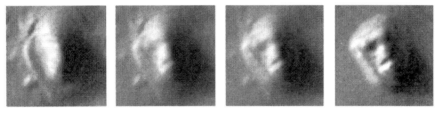

Figure 25 Similarity in structure between a rounded formation in the southwestern portion of the City and the Face. By overlaying the image of this formation over that of the Face a similarity in the gross morphology of the two formations is evident. The similarity is depicted here by fading from the rounded formation (left) to the Face (right). The transition from a featureless form to a face suggests the possibility that the Face might have been carved from a pre-existing landform.

7) Similarity Between the Face and a Rounded Formation — As noted earlier, the Face appears to rest on a highly symmetrical platform. A rounded formation located at the extreme southwestern end of the City also seems to rest on a similarly-shaped platform oriented in the same general direction.

8) Geometry of the Fort — In stark contrast to the sculpted appearance of the Face, located a short distance away, the Fort is a geometrically-shaped object with straight sides and sharp corners. Four straight sides or walls are visible in the two available images of this object. Hoagland noted that these walls seem to enclose an inner space; i.e., an area that is lower in height than the surrounding walls.

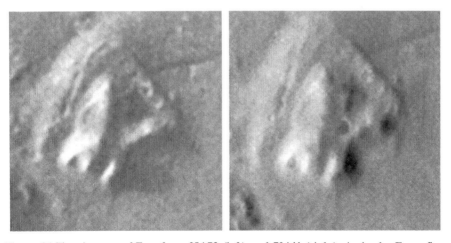

Figure 26 Two images of Fort from 35A72 (left) and 70A11 (right). As in the Face, fine-scale details can be seen in both images.

9) Fine-Scale Detail within the Fort — Like the Face, the Fort also contains subtle details that are at, or slightly below, the resolution of the imagery. Two of the walls have regularly-spaced indentations.

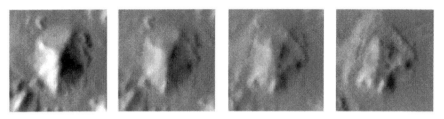

Figure 27 Similarity in structure between the Fort and adjacent 'Starfish' Pyramid from image 70A11. The sequence of images fades from the pyramid (left) to the Fort (right).

10) Similarity Between Fort and Adjacent Pyramidal Object — Another object of interest within the City is a formation known as the 'Starfish' Pyramid, which is next to the Fort. This object appears to be aligned with the Fort, and is similar in size and shape. 3-D visualization studies suggests the possibility that the Fort might have been a structure like the Starfish Pyramid that collapsed inward.

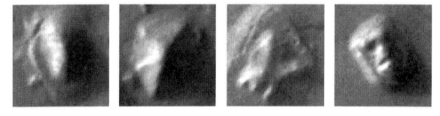

Figure 28 Similarity in orientation and scale of anomalous objects in frame 35A72. As in previous figures, these images have not been rotated or scaled in size.

Table 1 The orientations of the best defined edge on each of the above objects. Each angle is the average of three separate measurements. Angles are measured counter-clockwise from the horizontal axis (due east). The average (standard deviation) of the four measurements is 121.8° (1.6°). These objects appear to be very closely aligned to the Crater-McDaniel grid.

Measurement	Orientation
Left edge of Face	120.9°
Right edge of Fort	124.5°
Left edge of pyramid in City	120.8°
Left edge of rounded formation in City	120.8°

11) Similar Orientation of Fort, Face, Rounded Formation and Starfish Pyramid — Another early Mars investigator Daniel Drasin noted that many of the objects of interest in Cydonia are about the same size. In addition, the Face, rounded formation, Fort, and Starfish Pyramid all seem to be oriented in the same general direction (Table 1). This is particularly interesting because these objects are not all immediately adjacent to one another and do not have the same geomorphology.

12) Small Mound-Like Objects in City Arranged in Rectilinear Grid — Concerning the non-random distribution of the mounds analyzed by Crater and McDaniel, they have also shown that a subset of the mounds (A, B, D, E, and G) are arrayed in a grid-like pattern with a long/short side interval ratio of $\sqrt{2}$. The distances between the five mounds can be expressed as multiples of $\sqrt{2}$ and $\sqrt{3}$ times the distance between mounds B and D. All 30 angles between the five mounds can be expressed as $n(\pi/4)\pm m(t/2)$, where m and n equal 0, 1, 2, and 3, and $t = \arcsin(1/3) \approx 19.5°$. Measurements performed on map-projected Viking imagery reveal that the mounds are aligned in two directions: 34.53° (±0.91°) and 124.35° (±1.15°) — directions that differ by about 90°.

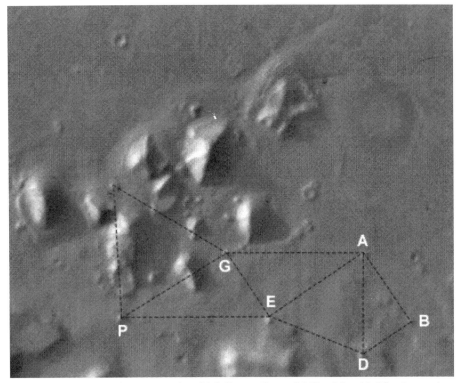

Figure 29 A subset of the mounds which lie on Crater-McDaniel grid. The image has been mapped to a Mercator projection.

13) Alignment of Objects in City and Face with the Grid Pattern — The Face, Fort, Starfish Pyramid, and rounded formation also appear to be oriented in the same general direction as the grid pattern established by the arrangement of the mounds. The similarity in orientation between the small mounds, three of the larger objects in the City, and the Face suggests an underlying regularity or pattern of organization in this collection of objects — a regularity that is hard to explain by geological processes alone.

Table 2 Orientations of the lines between the mounds. All measurements were performed on map-projected Viking images. Each measured value in the table is the average of three separate measurements. The average (standard deviation) of the first three measurements is 34.53° (0.91°), and of the fourth and fifth measurements is 124.35° (1.15°). The difference is close to 90° which suggests the presence of an underlying rectilinear pattern.

Measurement	Orientation
P to G	32.7°
E to A	35.9°
D to B	35.0°
E to G	123.2°
B to A	125.5°

14) D&M Pyramid — As noted earlier in this chapter, Torun could not find any plausible geological mechanism to account for the unusual shape of the D&M Pyramid. Using map-projected Viking imagery, it has also been determined that the south-facing side of the D&M Pyramid is oriented due south.

15) Anomalous Geomorphology of the Tholus — The Tholus is a large mound-like feature, not unlike Silbury Hill in England, located about 30 km southeast of the Face. It was originally discovered by Hoagland in seeking additional objects from which to derive spatial and angular relationships. According to geologists, James Erjavec and Ronald Nicks: "They [the Tholus and several other features with similar morphologies] display no evidence of vulcanism, are not impact related and are difficult to explain as remnants of larger non-distinct landforms because of their symmetrical shapes and uniform low gradients" [11].

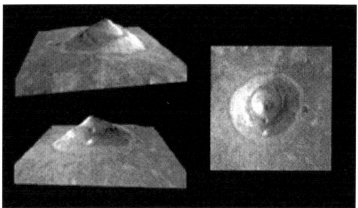

Figure 30 Overhead view of the Tholus from 35A74 (right) along with two perspective views generated from the northwest (top left) and southeast (bottom right).

16) Fine-Scale Detail on the Tholus — Like the Face and Fort, the Tholus also contains fine-scale details — details that should not be there if it is a natural object [2]. Two narrow grooves wind up the formation, clockwise and counter-clockwise, from the northwest to the southeast. On the southeast side there is a circular pit that can be clearly resolved in all three available images of the Tholus. 3-D analysis indicates that one set of grooves leads into this pit (perhaps an opening?) which is located about halfway up the side of the object.

17) The Cliff — This object is an elongated mesa topped by a sharp ridge-like feature running down its length. Hoagland, who discovered the Cliff, found that it lies roughly in line with the City square and the Face.

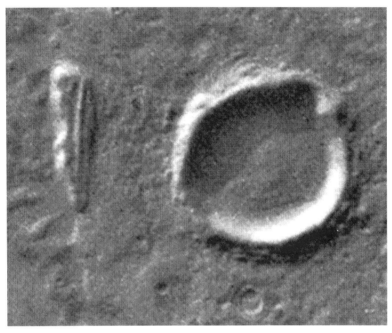

Figure 31 The Cliff and an adjacent impact crater. If the crater was caused by meteoric impact and the Cliff is a naturally occurring formation that existed prior to the impact as suggested by JPL, why are there no signs of damage or a debris flow around the Cliff as a result of the impact? (NASA/JPL)

Extraordinary Claims Require Extraordinary Evidence

Carl Sagan's *Parade* article critical of the independent Mars investigation ended with his famous saying: "Extraordinary claims require extraordinary evidence." For Sagan, extraordinary evidence meant there had be obvious indications of artificiality, like the signs of intelligent life he and Wallace found in their study of terrestrial satellite imagery. Because we had not found a single piece of extraordinary evidence, the 'smoking gun' as it were, the 'case' for artificiality was not credible, as far as Sagan was concerned. But when a smoking gun cannot be found, judgments in a court of law are made on circumstantial evidence — on the testimony of witnesses that point to the same conclusion. Science operates in a similar way. Take, for example, David McKay's 1996 paper concerning the discovery of possible fossilized micro-organisms in a meteorite thought to be from Mars [12]. McKay and his colleagues did not find a live microbe. Instead, they present five pieces of indirect evidence supporting their case for microbial life and state:

> None of these observations is in itself conclusive for the existence of past life. Although there are alternative explanations for each of these phenomena taken individually, when they are considered collectively, particularly in view of their spatial association, we conclude that they are evidence for primitive life on early Mars.

In effect the case for artificiality is based on the same kind of argument. Like the one for Martian microfossils, no single piece of evidence is sufficient in itself to prove these objects on Mars are artificial. But collectively, they all seem to point to the same conclusion.

If we were to bet on these objects being either natural or artificial, what are the odds? Mathematically, 'odds' is the probability that something is true given a piece of evidence divided by the probability that it is false given the same evidence. Fractal analysis provides one measure, one piece of evidence, for artificiality. If the fractal model-fit error is greater than a certain value an object is considered to be artificial; otherwise, it is natural. By testing the fractal technique on Earth, where we know the 'ground truth', we can determine the odds the fractal technique detects an artificial object. For scenes where the terrain is similar to that on Mars, the odds are between 3 to 1 and 5 to 1.

According to the above tally, there are 17 pieces of evidence in support of artificiality. Some are probably more reliable than the fractal technique, while others are less reliable. For example, finding a humanoid face in a terrestrial landscape is probably a stronger indicator of artificiality than finding several objects arranged in the same general direction. If we assume all pieces of evidence are independent (i.e., one does not depend on the other), and give the same odds to each, say 4 to 1, the odds for artificiality given all of the evidence is 4 raised to the 17th power, 4^{17}, or about 17 billion to one. But to be fair, since we are making an extraordinary claim, the likelihood that it is true *a priori*, from the outset, is very small, say a million to one against. So dividing the weight of the evidence by that amount gives a likelihood of about 17,000 to one in favor of artificiality[9].

What this says is that it is the quantity and diversity of all of the evidence together, rather than any one piece, that makes the case for artificiality so strong. The alternative hypothesis is, of course, that the Face and other objects

---

[9] Thomas Van Flandern has also performed this type of analysis. See his web site at http://www.metaresearch.org/solar%20system/cydonia/mrb_cydonia/new-evidence.asp

are simply naturally-occurring geological formations. But the odds of so many geological anomalies occurring in one place becomes exceedingly small, according to our numbers, only about 1 in 17,000. Although it is possible to use geological arguments to dismiss these anomalies individually, no consistent natural explanation has been put forth that is capable of explaining the diversity of forms, pattern of organization, and subtleties in design exhibited *in toto* by this collection of objects.

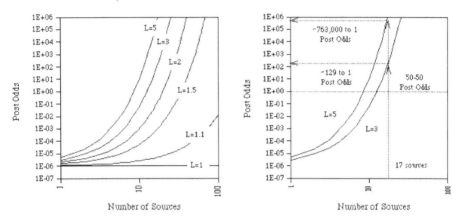

Figure 32 Post-odds increases dramatically as the number of sources increases for likelihood ratios greater than one (left). For N=17 sources and assuming individual source likelihood ratios between 3 and 5, the post odds are between 129 to 1 and 763,000 to 1 in favor of our hypothesis (right).

Sources

1. Carl Sagan, "The man in the moon," *Parade Magazine*, June 2, 1985.
2. M. J. Carlotto, *The Martian Enigmas: A Closer Look*, North Atlantic Books, Berkeley CA, Second Edition, 1997.
3. M. J. Carlotto, "Digital imagery analysis of unusual Martian surface features," *Applied Optics*, Vol. 27, pp 1926-1933, 1988.
4. C. Sagan, "The recognition of extraterrestrial intelligence," *Proceedings of the Royal Society*, Vol. 189, pp 143-153, 1975.
5. M. J. Carlotto and M. C. Stein, "A method for searching for artificial objects on planetary surfaces," *Journal of the British Interplanetary Society*, Vol. 43, pp 209-216, 1990.
6. B. O'Leary, "Analysis of images of the Face on Mars and possible intelligent origin," *Journal of the British Interplanetary Society*, Vol. 43, pp 203-208, 1990.
7. S. V. McDaniel, *The McDaniel Report*, North Atlantic Books, Berkeley CA, 1994.

8. R. Hoagland, *The Monuments of Mars: A City on the Edge of Forever*, North Atlantic Books, Berkeley CA, Second edition, 1992.
9. H.W. Crater and S.V. McDaniel, "Mound configurations on the Martian Cydonia plain," *Journal of Scientific Exploration*, Vol. 13, No. 3, 1999.
10. M. J. Carlotto, "Evidence in Support of the Hypothesis that Certain Objects on Mars are Artificial in Origin," *Journal of Scientific Exploration*, Vol. 11, No. 2, 1997.
11. J. Erjavec and R. Nicks, "A Geologic/Geomorphic Investigative Approach to Some of the Enigmatic Landforms in Cydonia," in *The Martian Enigmas*, (2nd Edition), North Atlantic Books, Berkeley CA, 1997.
12. D.S. McKay, E.K. Gibson, K.L. Thomas-Keprta, H. Vali, C.S. Romanek, S.J. Clemett, X.D.F. Chillier, C.R. Meachling, and R.N. Zare, "Search for past life on Mars: Possible relic biogenic activity in Martian meteorite ALH84001," *Science*, Vol. 273, 16 August 1996.

Six — City by the Sea

> They had a house of crystal pillars on the planet Mars by the edge of an empty sea, and every morning you could see Mrs. K eating the golden fruits that grew from the crystal walls... Afternoons, when the fossil sea was warm and motionless, ... you could see Mr. K himself in his room, reading from a metal book with raised hieroglyphs over which he brushed his hand, as one might play a harp. And from the book, as his fingers stroked, a voice sang, a soft ancient voice, which told tales of when the sea was red steam on the shore and ancient men had carried clouds of metal insects and electric spiders into battle. — Ray Bradbury, *The Martian Chronicles*

If there are ancient ruins on Mars, then there must have been intelligent life there at one time. If life was native to the planet as Brandenburg first suggested, then there had to have been enough time for it to begin and evolve — as it is thought to have evolved on Earth. Others, like Hoagland, have suggested the possibility that the City and Face were built by extraterrestrials, perhaps at a time when Mars was more habitable. Another possibility is that extraterrestrials might have terraformed the planet — altered the atmosphere and surface of Mars to be more Earth-like.

In the previous chapter, we established a consistent pattern of anomalies in Cydonia that suggest the City and Face might be artificial. But for this claim to be plausible, we must establish that life could have existed on Mars, if only for a relatively brief period.

The Crustal Dichotomy

The surface of Mars is divided into two distinct kinds of terrain (Figure 33). To the south are heavily cratered highlands, which cover about two thirds of the planet. The highlands are over 2 km in height, on average, and are thought to represent the oldest terrain on Mars. To the north are sparsely cratered plains that lie about 1 km below 'sea level' on average. The flatness of the plains and the lack of craters suggest this terrain is much younger than the more rugged highlands to the south. The transition or boundary between these two regions is marked by a series of cliffs, faults, and escarpments that lie roughly along a great circle inclined 35° to the equator. This division of the surface of Mars into smooth plains and rough highlands is known as the crustal dichotomy.

The origin of the crustal dichotomy is a mystery. A number of theories have been proposed to explain it. One is that the northern plains are the result of a gigantic impact event, which left a vast basin covering the top third of the planet. The northern plains would be the smooth center of the basin with the highlands starting at the periphery. A problem with this theory is that the concentration of mass deposited by the impact of a large meteor should be detectable as a gravity anomaly. Many such mass concentrations or 'mascons' have been found on the Moon under the lunar *maria*, which are thought be have formed from lava flows resulting from meteor impacts. No such mascons have been detected under the northern plains on Mars.

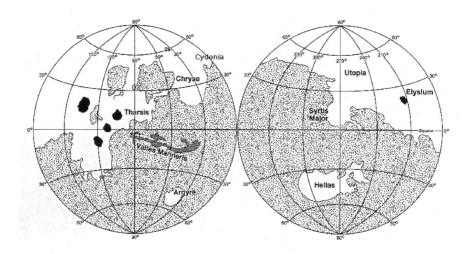

Figure 33 A schematic representation of Mars illustrating the division of its surface into smooth plains to the north (white) and heavily cratered highlands to the south (textured). Cydonia is located on the border between the two regions. (NASA)

Another possibility is that a series of smaller impacts coming from the same direction could be responsible for the dichotomy. According to Cattermole [1]:

> The practical problem with this idea is that it is difficult to envisage why concentration of such large basins should be so effectively polarized into one third of the surface area. Furthermore, there seems no getting away from the fact that if several overlapping basins had existed, regions of topography higher than the ancient surface must have been produced. There is no clear evidence of this.

Thomas Van Flandern suggests that instead of lowering the northern plains, debris from an exploded planet formed what is now the southern highlands. He believes that the material from the blast accumulated on the side that faced the explosion, while other side was shadowed, protected by the mass of the planet. His hypothesis, known as the Exploded Planet Hypothesis [2], is consistent with recent measurements of the topography and gravity of Mars by the Global Surveyor spacecraft

Another explanation for the dichotomy is that erosion was responsible for lowering the northern plains. However, studies have shown that no more than perhaps a few hundred meters of material could have been removed — not enough to explain the dichotomy. Some believe the dichotomy is neither ancient nor stable. Cattermole notes [1]:

Today the dichotomy zone appears to be a zone of active erosion, where scarp retreat is eating into the older cratered terrain. Such activity must have been going on for a considerable time, as is shown by the extensive occurrence on the plains side of the dichotomy of knobby terrain — consisting of isolated knobs and small mesas that appear to be relics of the old plateau.

The City and Face are located along this zone of active erosion. They too are considered remnants of the older terrain. According to Michael Malin[10]:

> The Cydonia region lies on the boundary between ancient upland topography and low-lying plains, with the isolated hills representing remnants of the uplands that once covered the low-lying area. The features [the City and Face] seen in these mesas and buttes (to bring terrestrial terminology from the desert southwest to bear on the problem) result from differential weathering and erosion of layers within the rock materials.

The basic idea is that this entire area was once covered by several kilometers of soft sediment, which was striped away by wind and other erosive forces, leaving the more durable knobs and mesas we see today. Although this explanation seems reasonable at first, it does not hold up under scrutiny. First, it is too simple. Geological observations based on Viking imagery indicate that multiple periods of erosion and deposition have taken place. The geological record of Cydonia is much more complicated than Malin's explanation leads us to believe. Second, where did all of the eroded material go? Geologist James Erjavec puts it this way [3]:

> [If] a one to two kilometer overburden once covered this area, a very large volume of sediment would have to have been removed and deposited somewhere. Since Mars of today has no liquid oceans or large bodies of water, there is no place for that sediment to 'hide'. The material should be identifiable somewhere on the surface of Mars. The reality is that there appears to be no place on Mars where such a large volume of sediment has been deposited.

A Paleo-Ocean on Mars?

Hoagland and the independent Mars investigation team first noticed that the City and Face were located near the 0 km datum, which suggested the possibility that the site was near the shoreline of an ancient body of water. Their poster paper, "Preliminary Report of the Independent Mars Investigation Team: New Evidence of Prior Habitation?" presented at the 1984 *Case for Mars II* conference, was the first to suggest the possibility that the northern plains "was the sedimentary deposit of a former ancient ocean, which would have occupied essentially the entire northern hemisphere of Mars" [4].

---

[10] http://www.skepticfiles.org/skep2/marsobs1.htm

They also observed large polygonal cracks in what many planetary scientists now believe is a thick layer of sediment in the northern plains. On Earth, similar cracks are found near glaciers. Glacial activity is also suggested in patterns of curvilinear ridges present in the northern plains. Other features found on Mars, which on Earth are indicative of ground ice and permafrost, include pseudo craters, which are produced by the explosive release of groundwater heated by lava, pingos — dome-like mounds with ice core interiors, and patterned terrain such as circles, polygons, nets, and stripes created by ice. All of these features are found in the immediate vicinity of the anomalies in Cydonia.

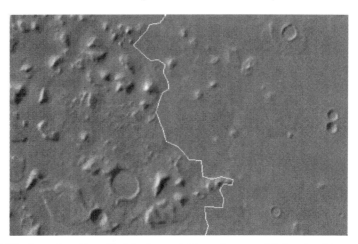

Figure 34 The boundary identified by James Erjavec between higher, knobby terrain and a lower pediment surface that may have been an ancient shoreline in Cydonia.

Erjavec performed a detailed study of the distribution of impact craters in the knobby terrain — isolated knobs and small mesas that appear to be relics of the old plateau, and the smoother sediment-covered pediment surface. In Cydonia, the knobby terrain lies to the west and includes the formations in the City (Figure 34). To the east is the pediment surface containing the Face. Erjavec found that both regions have a similar distribution of large impacts (greater than 1 km in diameter), but the pediment surface has a significantly greater number of smaller craters (less than 1 km in diameter). This difference in crater counts implies that a single mechanism such as differential erosion could not have been responsible for the formation of these objects in this region of Mars.

The morphology of the objects in the City differs markedly from the Face. Where the Face has a smoothed appearance, the objects in the City are angular in shape. One mechanism that could have acted selectively on these objects and the surrounding terrain is water. The idea is that the smaller impact craters on the pediment surface were preserved under water, while those above water, in the knobby terrain, were being erased by erosion. If the Face was near the shoreline, perhaps an island, as its base was being smoothed by wave action, the angularity of the larger structures on dry land, in the knobby terrain, would be unaffected.

In support of this theory Erjavec and others have identified possible shoreline features in the area including strand lines — narrow linear features that are often visible along the edges of dry lake beds that mark stable water levels over time, wave cut benches, shore platforms — sloped inter-tidal rock surfaces, and others. Erjavec also found what appear to be landslides from knobs extending into the pediment surface. These could be interpreted as evidence of shoreline erosion, the landslides a result of wave action undercutting the knobs.

In a paper entitled "Geomorphology of Selected Massifs On the Plains of Cydonia, Mars" published in the summer of 1999, David Pieri further advances the idea of an ancient ocean in Cydonia. His assessment of the Face and other similar landforms (which he refers to as massifs), based on the more recent Mars Global Surveyor (MGS) imagery, is this [5]:

> [T]aken as a group, these features appear to have at least a qualitative morphological affinity with terrestrial submarine or lacustrine features. The (a) softened, relatively non-angular appearance, (b) the evidence of substantial mass wasting, (c) the lack of any apparent rilling even in high spatial resolution images, (d) draped morphology of superjacent thin sedimentary layers, and (e) the shelf-like morphology of the basal geologic layer are all qualities consistent with these massifs having been at least modified in a subaqueous (as opposed to subaerial) environment, that could have been marine, lacustrine, or possibly littoral (i.e., near shore). While individual massifs may exhibit some or all of these qualities, a more compelling case is made when the isolated massifs are assessed as a group, in which the described morphologies are repetitive and reinforcing.

In this article, Pieri argues the Face is a natural feature. He states that a lack of rilling on these massifs indicates that the primary morphologies were developed in a submarine or lacustrine environment. In response to Pieri, Erjavec and others counter [6]:

> Erjavec and Brandenburg have found what appear to be rills on several Cydonian landforms. This is strong evidence that this area was aerially exposed and erosion occurred through the actions of both precipitation and surface runoff. In combination with the lacustrine or marine signature of this area, it strongly suggests that the morphologies of the Cydonian massifs are polygenetic in origin. This is an important point as it implies that the 'Face' was exposed during a time when Mars still had a hydrogeologic cycle.

In the next chapter we will return to the debate concerning the origin of the Face. But for now the growing body of evidence suggesting that a large body of water once existed in Mars raises the obvious question: Where did all that water go?

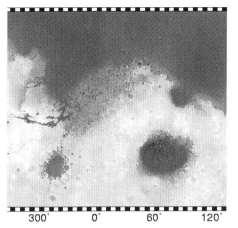
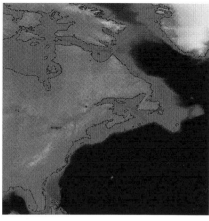

Figure 35 A portion of a global MOLA[11]-derived elevation map enhanced to better show possible shoreline features (left). If Mars once contained a large northern ocean, these features might indicate different levels of the ocean. A portion of a 5-arc minute world digital elevation model showing the eastern seaboard of North America (right). The black line is the current shoreline. Data have been enhanced to show the continental shelf and channels that are today below sea level. (NASA/JPL/USGS)

Mars Past, Present, and Future

Today Mars' atmosphere is extremely thin, consisting primarily of carbon dioxide plus small amounts of water vapor. Temperatures average around -60° Celsius (-76° Fahrenheit). In the summer, afternoon temperatures can reach as high as 20° C (68° F) at the equator. Polar nights can get as cold as -120° C, well below the freezing point of CO_2. Although daytime temperatures can melt ice, most scientists believe that liquid water cannot exist on the surface because the atmospheric pressure is too low. Like dry ice on Earth, water ice on Mars that is heated by the sun sublimates directly into vapor.

Despite current conditions, it is certain that large quantities of water and ice once existed on Mars. Mariner 9 photographed numerous channels from orbit ranging from broad sinuous channels to narrow dendritic channel networks. Some of the largest channels, 30-60 km wide and up to 1200 km long, originate in the chaotic terrain and canyons north of Valles Marineris along the equator. Following the slope of the terrain, the channels flow northward into Chryse Planitia, which is southwest of Cydonia. According to a NASA fact sheet, "These channels resemble features produced by episodic floods on Earth... Catastrophic melting of ground ice could form both the chaotic terrain and the giant flood channels in a single event."

---

[11] The Mars Orbiter Laser Altimeter (MOLA) is an instrument on board MGS.

In 1997, Mars Pathfinder landed in one of these outwash channels known as Ares Valles (Figure 37). Pathfinder photographs provide further evidence of massive flooding on Mars. According to JPL, "Boulder trains seen in distant views of the twin hills [a rock formation photographed by the Pathfinder camera] resemble landforms found in the lee of obstacles in large terrestrial floods. The ridge-and-valley topography is generally consistent with fluvial landforms developed in and comprised of mobile sedimentary material" [7]. They estimate the size of the flood to have been comparable to large floods that have occurred in Iceland and Washington state.

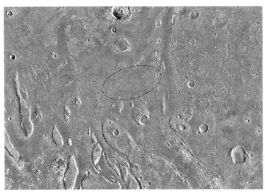

Figure 36 Ares Vallis, a large outwash plain near Chryse Planitia. This region is one of the largest outflow channels on Mars, the result of a huge flood over a short period flowing into the Martian northern lowlands. The ellipse gives the general location of the Mars Pathfinder landing site. (NASA/JPL)

Figure 37 Panorama of Ares Valles from the surface, photographed by the Mars Pathfinder. (NASA/JPL)

Sojourner, a telerobotically-controlled rover dispatched from Pathfinder, discovered small pebbles on the surface and in rocks. Rover scientists believe the pebbles could have been "liberated from sedimentary rocks composed of cemented silts, sands, and rounded fragments; such rocks are called conglomerates." They go on to say that "pebbles in conglomerates would suggest that liquid water existed at the surface before the Ares floods" [8].

Figure 38 Close-up photograph of a rock near the Pathfinder landing site photographed by a camera on the telerobotically-controlled Rover. Arrows indicate what appear to be small pebbles in the rock. (NASA/JPL)

For sedimentary rocks to have formed, large standing bodies of water must have existed on Mars for a considerable period. But where did this water come from? One source could have been the rapid melting of ground ice (permafrost) triggered by some sudden event. Atmospheric precipitation in the form of rain and snow is another possibility. In addition to the wide outwash channels discussed above, narrow valleys — some with tributaries forming dendritic drainage patterns — were also imaged by Mariner 9 and Viking Orbiter. Their lack of an apparent source and characteristic pattern strongly suggests, by comparison with terrestrial features, that they were formed by precipitation and runoff. There is also evidence to suggest that some Martian valleys might have been shaped by glaciers.

For large quantities of water to have existed on Mars in the past, the atmosphere would have had to have been much thicker. At present, the atmospheric pressure on Mars is about 8 millibars. Compared to our standard sea level pressure of 1013 millibars, Mars' atmosphere is less than 1/100th that of Earth. An atmospheric pressure of 2-5 bars (2000-5000 millibars) of carbon dioxide was required to raise the annual average temperature on Mars to the melting point of water ice billions of years ago, when the sun was dimmer than it is today. Over time, as the luminosity of the sun increased to its present brightness, lesser amounts, only about 500-1000 millibars, would be required.

A planet's surface temperature depends on its distance from the sun, the composition of its atmosphere, which determines how much heat is retained, and the amount of internal heat generated by the planet. Carbon dioxide, CO_2, is one of several 'greenhouse gases' that allow visible light to pass but blocks infrared radiation (heat). The amount of heat retained by a planet depends largely on the amount of CO_2 gas in the atmosphere. The current atmosphere of Mars retains very little heat. Venus' thick CO_2 atmosphere keeps almost all of its heat, with surface temperatures close to 900° F. The Earth's atmosphere contains much less CO_2 than that of Venus, so the greenhouse effect is weaker. However, without CO_2 in our atmosphere, the average global temperature would be well below freezing, and Earth would be unable to support life as we know it, much like Mars[12].

As on Earth and Venus, volcanoes on Mars are believed to have generated a thick CO_2 atmosphere early in the planet's history. Although some of Mars' atmosphere has escaped into space, much of it is thought to be 'trapped' in rock. CO_2 reacts with water to form carbonic acid (soda water), which in turn, reacts with various minerals in rock to produce clays and carbonates. The theory is that, over time, as atmospheric CO_2 was absorbed by large standing bodies of water on Mars, thick layers of carbonate-rich sediment accumulated and eventually consolidated to form sedimentary rock. This model is supported by Pathfinder images that show evidence of conglomerates — sedimentary rocks composed of cemented silts, sands, and pebbles[13].

Some believe Mars lost its initial CO_2 atmosphere to carbonate formation in as few as several tens of millions of years. This view is consistent with the apparent lack of plate tectonics on Mars. Earth's crust is divided into plates. As our plates shift and slide underneath one another (a process known as subduction), sedimentary deposits rich in atmospherically- and biogenically-derived carbonates (on Earth, the skeletal remains of dead marine creatures) are dragged under the surface and heated, freeing trapped CO_2, which is returned to the atmosphere along with other gases through volcanic eruptions. In effect, plate tectonics recycle Earth's CO_2. Evidence suggests that much of the Martian surface has not changed very much over time. Craters, supposedly formed billions of years ago during the 'heavy bombardment' — an early phase in the formation of the solar system, are still visible, and enormous volcanoes, which on Earth would be worn down, subducted, and replaced by new ones, remain. The lack of change indicates that tectonic processes do not occur on Mars and so there is no mechanism to recycle Martian CO_2. As Mars lost its atmosphere and cooled, it became the planet we see today.

Yet, the Pathfinder results seem to indicate subsequent events ripped these layers of sediment apart and transported them considerable distances. Mariner 9 imagery provided visual evidence of episodic floods on Mars. Vic Baker and his colleagues at the University of Arizona developed a climatological model to explain them [9]:

---

[12] At present, Earth faces the opposite problem of global warming with too much carbon dioxide in the atmosphere retaining too much heat.

[13] In addition to carbon dioxide, early Mars may have had sulfur dioxide in its atmosphere as well, which would have interfered with the formation of carbonates. This could explain the lack of carbonates on Mars today.

The formation of valley networks early in Mars' history is evidence for a long term hydrological cycle, which may have been associated with the existence of a persistent ocean. Cataclysmic flooding, triggered by extensive Tharsis volcanism, subsequently led to repeated ocean formation and then dissipation on the northern plains, and associated glaciation in the southern highlands until relatively late in Martian history.

According to their model, a cycle of climate change begins with an increase in volcanic activity. The associated heat flow warms ground water and triggers outburst flooding from reservoirs of ground ice and water forming the chaotic terrain that we see today. The resulting flood of water reaches the northern plains and releases water vapor through evaporation. As the north plains fill with water, CO_2 in the north polar cap is released into the atmosphere along with CO_2 trapped in rock and ground water. The warming greenhouse atmosphere precipitates snow to form glaciers in the south polar regions and possibly rain to the north. As water seeps back into the ground and CO_2 again becomes locked up as carbonates in rock and dry ice at the polar caps, the atmosphere thins and the planet cools. Baker believes that this cycle has occurred many times on Mars, and concludes by saying "Over the history of the planet, the declining internal heat flow, atmospheric escape of hydrogen, and sequestering of CO_2 in carbonates and H_2O in clays would gradually reduce the activity in successive cycles." They believe the most recent cycle ended a few hundred million years ago.

It is conceivable that the City and Face might have been built during one of these warmer and wetter periods on Mars. But in theory, they could have been built at any time. Chris McKay and others have suggested the possibility of terraforming Mars using existing technology applied on a planetary scale [10]. The scenario starts with the warming of the Martian poles using mirrors or black soot to absorb sunlight. This, in turn, releases CO_2 into the atmosphere. As the planet warms, additional CO_2 and water vapor are released from the soil adding to the greenhouse effect, which further increases the temperature. This initial phase of terraforming might take perhaps 100 years. As the CO_2 atmosphere thickens, plants can be introduced to begin to generate oxygen by photosynthesis. The second phase of terraforming would take much longer; McKay estimates over 100,000 years would be required to produce a breathable atmosphere.

It should be noted that the red color of Mars is due to the presence of iron oxides (rust) on the surface. The presence of oxides strongly suggests that an oxygen-rich atmosphere once existed on Mars. Earth's oxygen is the result of photosynthesis. By analogy, it is possible that if Mars had an atmosphere high in oxygen, it too could have been the result of widespread photosynthesis from plant life on the surface. The technical feasibility of terraforming Mars today suggests the possibility that it might have been more Earth-like in its past.

Having established Mars could have once been a much warmer and wetter place leads to the next question: Could it have supported life?

Evidence of Life on Mars

The possibility that primitive life (microbes) might have existed billions of years ago on Mars is widely accepted by planetary scientists today. According to Chris McKay, "From a biological perspective, the most important information returned from spacecraft exploration of Mars may be the geological evidence that liquid water was abundant on the Martian surface at some time in the past. Water is the quintessence of life — all life on Earth requires the presence of liquid water. Liquid water is the only compound for which this statement is true" [11].

There is a growing body of circumstantial evidence that suggests microbial life existed on Mars, and may still exist today. In 1975, two Viking spacecraft, each with an orbiter and a lander, were sent to Mars to look for evidence of life on the Martian surface. Each of the two landers contained three biology experiments to test for the presence of microbial life in the Martian soil [12]. In the first experiment (PR), nutrients were added to a soil sample, which was incubated under a light in the presence of radioactively labeled carbon monoxide and carbon dioxide. After incubation the soil was heated to see if any of the labeled carbon was metabolized to form organic molecules. In the second experiment (GEX), changes in the composition of gases in contact with the soil and nutrients were monitored. In the third experiment (LR), a labeled nutrient was added to the soil and the gases monitored to see if the nutrient broke down. In the PR experiment a small amount of carbon was converted into organic molecules, in the GEX experiment a small amount of carbon dioxide, nitrogen, and other gases were produced, and in the LR experiment there was an initial rapid release of labeled gas followed by a slower release. All were positive results. However, conflicting evidence was provided by the Gas Chromatograph Mass Spectrometer (GCMS) experiment, which found no organic compounds on Mars. According to Gil Levin, a member of the life science team, "This major dilemma was 'solved' by some members of the scientific community by opting for the conservative conclusion, chemistry. Various extreme environmental factors were cited to support the view that Mars is hostile to life." But Levin disputes this assessment [13]:

> The Viking Labeled Release (LR) experiments conducted on Mars in 1976 returned data satisfying its pre-mission criteria for the presence of life in the samples of surface material analyzed... Both Viking landing sites, 4,000 miles apart, produced strong positive responses. As controls, portions of the samples which had produced the positives were heated to temperatures designed to distinguish chemical from biological agents. Each of a total of nine separate reactions was consonant with a biological entity. Together, the results constitute strong evidence for the existence of microorganisms on Mars.

Although Levin's interpretation is not widely accepted within the planetary community, the facts remain.

Evidence of past life on Mars comes from a collection of meteorites known as SNC meteorites, for the three distinct classes of meteorites found in Shergotty, India, Nakhla, Egypt, and Chassigny, France. By comparing the composition of gases trapped within the meteorites to that of the Martian atmosphere measured by the Viking landers, the SNC meteorites have been determined to be Martian in origin.

Although it is not technically a SNC meteorite, ALH84001, which was discovered in Antarctica, is believed to have been blasted from Mars by a huge impact about 16 million years ago. ALH84001 is as old as Mars itself — about 4.5 billion years. David S. McKay and his colleagues summarize their analysis of the meteorite [14]:

> In examining the Martian meteorite ALH84001 we have found that the following evidence is compatible with the existence of past life on Mars: (i) an igneous Mars rock (of unknown geologic context) that was penetrated by a fluid along fractures and pore spaces, which then became the sites of secondary mineral formation and possible biogenic activity; (ii) formation age for the carbonate globules younger than the age of the igneous rock; (iii) SEM and TEM [electron microscope] images of carbonate globules and features resembling terrestrial microorganisms, terrestrial biogenic carbonate structures, or microfossils; (iv) magnetite and iron sulfide particles that could have resulted from oxidation and reduction reactions known to be important in terrestrial microbial systems; and (v) the presence of PAHs [polycyclic aromatic hydrocarbons] associated with surfaces rich in carbonate globules.

They go on to state: "None of these observations is in itself conclusive for the existence of past life. Although there are alternative explanations for each of these phenomena taken individually, when they are considered collectively, particularly in view of their spatial association, we conclude that they are evidence for primitive life on early Mars."

Seven years earlier, I.P. Wright, M.M. Grady, and C.T. Pillinger had published a paper "Organic Materials in a Martian Meteorite" in *Nature*, which was even much more thought provoking. "The meteorite EETA 79001 which many believe to have originated on Mars contains carbonate materials thought to be Martian weathering or alteration products. Accompanying the carbonates are unexpectedly high concentrations of organic materials..." [15]. Where ALH84001 was 4.5 billion years old, meteorite EETA 79001 had formed much more recently, about 200 million years ago.

Although Wright and his colleagues received little attention, David McKay's results, published in the August 16, 1996 issue of the journal *Science*, received a great deal of publicity, especially from NASA. Administrator Dan Goldin used the opportunity to make the following statement at a press conference:

> NASA has made a startling discovery that points to the possibility that a primitive form of microscopic life may have existed on Mars more than three billion years

ago. The research is based on a sophisticated examination of an ancient Martian meteorite that landed on Earth some 13,000 years ago.

The evidence is exciting, even compelling, but not conclusive. It is a discovery that demands further scientific investigation. NASA is ready to assist the process of rigorous scientific investigation and lively scientific debate that will follow this discovery.

I want everyone to understand that we are not talking about 'little green men'. These are extremely small, single-cell structures that somewhat resemble bacteria on Earth.

There is no evidence or suggestion that any higher life form ever existed on Mars.

Figure 39 Meteorite ALH84001 found in Antarctica in 1984 that is believed to have been blasted from Mars by an impact about 16 million years ago. (NASA)

A Theory of Life

In the 1920's, A. I. Oparin and J. B. S. Haldane postulated that Earth's early atmosphere, in conjunction with high levels of ultraviolet radiation from the sun, could enhance the chemical reactions required to produce organic molecules, and ultimately life. In 1953, Harold Urey and Stanley Miller demonstrated that amino acids, simple sugars, and nucleotides could be produced by passing electric currents through a mixture of methane, ammonia, hydrogen, and water. This result is widely accepted as proof of the Oparin-Haldane Chemical Evolution Hypothesis — that life formed abiotically from a chemical 'soup' early in Earth's history.

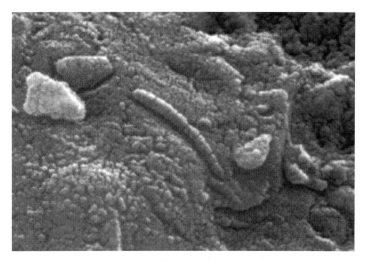

Figure 40 A high-resolution scanning electron microscope image that shows a tube-like object less than 1/100th the width of a human hair. This object is believed to be the fossilized evidence of primitive life that existed on Mars 3.6 billion years ago. (NASA)

It is thought that soon after Earth's crust solidified, about 4 billion years ago, primitive bacteria first appeared. These early cells, known as prokaryotes, had no nucleus. Two billion years later, nucleated cells known as eukaryotes appear in the fossil record. Although it is believed that eukaryotes 'evolved' from prokaryotes, there is no evidence to support it. By about 600 million years ago, evidence of multicellular eukaryotes can be found in the fossil record. Then, approximately 500 million years ago, there was an inexplicable increase in the number and diversity of life forms on Earth. During the so-called 'Cambrian Explosion', which lasted for 5-10 million years, all major groups of animal life (phyla) on Earth appeared. Equally inexplicable, assuming classical Darwinian evolution is correct, is that no additional phyla have appeared since that time.

Evidence from the SNC meteorites suggests that microbial life existed on Mars throughout its history. A growing body of evidence also points to repeated cycles of climatological change on Mars. In the absence of plate tectonics, these intermittent warm and wet periods could have lasted for as long as several tens of millions of years, until the atmosphere once again became trapped as carbonates in Martian rock. In the opinion of most planetary scientists, tens of millions of years is simply not enough time for advanced life — one capable of building the City and Face — to have developed on Mars.

Examination of the fossil record reveals the evolution of life on Earth was not a gradual process. Instead, it appears to have been punctuated by sudden appearances of new biological forms followed by long periods of little or no change in between. Almost as remarkable as the Cambrian Explosion is the rapid evolution of Homo sapiens from earlier species in only 3-4 million years.

Could advanced life forms elsewhere have developed in shorter, more compressed, timeframes?

Given the relatively short period of the Cambrian explosion and the evolution of Homo sapiens, it is not unreasonable to hypothesize that advanced life forms might have developed on Mars over a more compressed timeframe, perhaps early in the planet's history as first suggested by Brandenburg [16], or during one of its brief episodic periods of climate change? Of course it is also possible that extraterrestrials might have visited our solar system (a possibility first suggested by Carl Sagan in 1963) and established a semi-permanent presence on Mars, perhaps even terraforming it to suit their needs.

Sources

1. Peter Cattermole, *Mars, The Story of the Red Planet*," Chapman and Hall, London 1992.
2. Thomas Van Flandern, *Dark Matter, Missing Planets and New Comets*, North Atlantic Books, Berkeley CA, 1999.
3. *The Case for the Face, Scientists Examine the Evidence for Alien Artifacts on Mars*, Stanley V. McDaniel and Monica Rix Paxson, Editors, Adventures Unlimited Press, Kempton IL, 1998.
4. Richard C. Hoagland, "Preliminary Report of the Independent Mars Investigation Team: New Evidence of Prior Habitation?," in *The Face on Mars: Evidence for a Lost Civilization?*, by Randolpho Pozos, Chicago Review Press, Chicago IL, 1986.
5. David C. Pieri, "Geomorphology of Selected Massifs On the Plains of Cydonia, Mars," *J. Sci. Exploration*, Vol. 13, No. 3, 1999.
6. Mark J. Carlotto, Horace W. Crater, James L. Erjavec, and Stanley V. McDaniel, "Response to Geomorphology of Selected Massifs On the Plains of Cydonia, Mars by David Pieri," *J. Sci. Exploration*, Vol. 13, No. 3, 1999.
7. P. H. Smith, J. F. Bell III, N. T. Bridges, D. T. Britt, L. Gaddis, F. Gliem, R. Greeley, H. U. Keller, K. E. Herkenhoff, S. Hviid, R. Jaumann, J. R. Johnson, R. L. Kirk, M. Lemmon, J. N. Maki, M. C. Malin, S. L. Murchie, J. Oberst, T. J. Parker, R. J. Reid, P. Rueffer, R. Sablotny, L. A. Soderblom, C. Stoker, R. Sullivan, N. Thomas, M. G. Tomasko, W. Ward, E. Wegryn, "Results from the Mars Pathfinder Camera," *Science*, Vol. 278, 5 Dec 1997, pp. 1758 — 1765.
8. J. R. Matijevic, J. Crisp, D. B. Bickler, R. S. Banes, B. K. Cooper, H. J. Eisen, J. Gensler, A. Haldemann, F. Hartman, K. A. Jewett, L. H. Matthies, S. L. Laubach, A. H. Mishkin, J. C. Morrison, T. T. Nguyen, A. R. Sirota, H. W. Stone, S. Stride, L. F. Sword, J. A. Tarsala, A. D. Thompson, M. T. Wallace, R. Welch, E. Wellman, B. H. Wilcox, D. Ferguson, P. Jenkins, J.

Kolecki, G. A. Landis, D. Wilt, " Characterization of the Martian Surface Deposits by the Mars Pathfinder Rover," *Science*, Vol. 278, 5 Dec 1997, pp. 1765-1767.

9. V. R. Baker, R. G. Strom, V. C. Gulick, J. S. Kargel, G. Kamatsu, and V. S. Kale, "Ancient Oceans, Ice Sheets, and the Hydrological Cycle on Mars," *Nature*, Vol. 352, 15 Aug. 1991, pp. 589-594.

10. Christopher P. McKay, Owen B.Toon, and James F. Kasting, "Making Mars habitable," *Nature*, Vol. 352, 8 Aug. 1991, pp. 489-496.

11. C. P. McKay, R. L. Mancinelli, and C. R. Stoker, "The Possibility of Life on Mars During a Water Rich Period," in *Mars*, H. H. Kieffer, B. M. Jakosky, C. W. Snyder, M. S. Matthews (Editors), University of Arizona Press, Tucson AZ, 1992.

12. Harold P. Klein, "The Search For Life on Mars," in *The Surface of Mars* by Michael H. Carr, Yale University Press, New Haven CT, 1981.

13. Gilbert V. Levin, "The Viking Labeled Release Experiment and Life on Mars," *Proceedings of SPIE Instruments, Methods, and Missions for the Investigation of Extraterrestrial Microorganisms*, July-August 1997, San Diego, CA.

14. McKay, D.S., Gibson, E.K., Thomas-Keprta, K.L., Vali, H., Romanek, C.S., Clemett, S.J., Chillier, X.D.F., Meachling, C.R. and Zare, R.N., "Search for past life on Mars: Possible relic biogenic activity in Martian meteorite ALH84001," *Science*, Vol. 273, 16 August 1996.

15. I.P. Wright, M.M. Grady, and C.T. Pillinger, "Organic Materials in a Martian Meteorite," *Nature*, Vol. 340, 20 July 1989.

16. V. DiPietro, G. Molenaar, and J. Brandenburg, "The Cydonia Hypothesis," *J. Sci. Exploration*, Vol. 5, No. 1, 1991.

Seven — The Politics of Cydonia

> No part of the aim of normal science is to call forth new sorts of phenomena; indeed those that will not fit into the 'box' are often not seen at all. Nor do scientists normally aim to invent new theories, and they are often intolerant of those invented by others. — Thomas Kuhn, *The Structure of Scientific Revolutions*

The detection of possible archaeological ruins on Mars by the Viking orbiter was completely unexpected. As we saw earlier, speculation about intelligent life on Mars peaked late in the 19th century and then began to decline as telescopic observations and early planetary probes showed Mars to be a far less hospitable place than originally thought. Although Mariner 9, and later Viking painted a different picture of the Red Planet — one with enormous volcanoes, vast canyon systems, and channels that showed that Mars once had water — the consensus within the planetary science community was that a warm and wet environment did not last long enough for advanced life forms to develop as they had on Earth.

According to Kuhn, "In science...novelty emerges only with difficulty, manifested by resistance, against a background provided by expectation. Initially only the anticipated and usual are experienced, even under circumstances where anomaly is later to be observed" [1]. Without a doubt, the City and Face represent the greatest concentration of anomalous surface features discovered to date on another planet, and, as we saw in the previous chapter, that they lie in what was probably, at one time, a warm maritime environment compounds the anomaly.

That outsiders like ourselves were advancing a radical hypothesis — one that might completely overturn prevailing scientific theories about Mars — was not well-received by the planetary science community. Some ignored, and even ridiculed us. But in order to test our hypothesis, to either to confirm or deny it, additional high-resolution pictures of Mars were needed.

Mars Observer

Around the time DiPietro and Molenaar were announcing their initial findings about the Face on Mars, the Viking mission was coming to an end. The Viking 2 orbiter was powered down on July 25, 1978 after 706 orbits. Viking 1 followed two years later on August 17, 1980, after over 1400 orbits.

The next mission to Mars, the Mars Geoscience/Climatology Orbiter, was in the planning stages. In part, because Viking had been so successful in imaging the Martian surface, a camera was not originally included in the mission. Cost too was an issue. With its budget shrinking, in 1980, NASA's Solar System

Exploration Committee developed a twenty-year plan for planetary exploration. Where the cost of the Viking mission was over $1 billion, MG/CO, later renamed Mars Observer, was to be the first of a new breed of low-cost spacecraft containing instruments developed and operated by small university-led teams. Michael Malin, then at the University of Arizona, proposed a high-resolution camera be added to the Mars Observer spacecraft:

> Much of what is known about the state and evolution of the terrestrial planets comes from the study of images. This is particularly true for Mars... Nearly one half of the Viking Mars Science Highlights refer to imaging results, clearly demonstrating that even after three previous, imaging-intensive missions, significant discoveries remained to be made by cameras. Many of these discoveries resulted from the increase in spatial resolution or temporal coverage afforded by each, more sophisticated, follow-on mission [2].

In 1985, Malin's Mars Orbiter Camera became one of eight experiments selected to be carried aboard the Mars Observer. Originally scheduled to be launched by the Space Shuttle in August 1990, the mission was postponed to 1992 and modified for launch aboard a Titan 3 rocket after the Space Shuttle Challenger accident in 1986.

Meanwhile, the Soviets were preparing to send a pair of spacecraft to orbit Mars and rendezvous with its tiny moon Phobos. Previously, they had launched eight probes to Mars in the 1960's. All were failures. They were somewhat more successful in the 1970's, sending six spacecraft to the Red Planet, one of which became the first to land on Mars in 1971. In contrast to their highly successful Venera missions to the planet Venus, the Soviets continued to have problems with their Mars probes. Phobos 1 and 2 were both launched in July 1988. One was lost several months into the mission when ground controllers sent the wrong command to the spacecraft. The second entered orbit around Mars, but later failed due to a computer malfunction, after approaching to within several hundred kilometers of Phobos.

In 1990, Malin formed Malin Space Science Systems, a private company to design, develop, and operate instruments to fly on unmanned and manned spacecraft. Mars was its focus. Under the terms of its contract with JPL, in addition to building the camera, MSSS was responsible for operating it as well. While previous planetary missions had been controlled from JPL, the Mars Observer operations were distributed, with each team controlling its own instrument from their home institution. By means of a highly automated ground data system [3], MSSS would be able to control the camera using only a fraction of the staff that had been required in previous missions. As president of MSSS and the principal investigator for the MOC, Malin was in the position to control, almost single-handedly, all of the high-resolution imaging of the Martian surface. Moreover, according to the terms of his contract, Malin was not required to release the imagery to the public until after a six to nine month 'data validation' period.

To those of us interested in new pictures of the City and Face, it became clear that we needed to demonstrate to Malin the validity of our findings in order to get him to take those pictures. That this would be difficult became evident in the late 1980's.

In the spring of 1989, Hoagland asked Erol Torun and me to support a meeting he had set up with Representative Robert A. Roe, Chairman of the House Science, Space, and Technology Committee. Unable to make much of a direct impact on the planetary science community, Hoagland's strategy was to apply political pressure to NASA to get them to re-image Cydonia with the Mars Observer. We briefed Roe and his aid, Robert Maitlin, for over an hour on our research. By the end of the meeting both were convinced that new photographs were needed to resolve the controversy. In a letter to one of his constituents, Roe stated:

> NASA leadership has assured me of the following: The Mars Observer will hopefully be launched in 1992 to follow up discoveries of Mariner 9, Viking I and Viking II. The Mars Observer is equipped with both a wide-angle camera and a narrow field of view camera. The wide angle camera will be used to survey the entire surface of Mars. The narrow angle camera will be used to focus on selected points of interest whenever possible. It is my understanding that NASA does intend to try to capture, with the narrow angle camera, the Cydonia region including the unique features ... referred to as the 'pyramids' and the 'face'. NASA will attempt to locate and focus on this region both because it has interesting geological features and has attracted wide popular interest.

But then, in an apparent reaction to Roe's request to NASA, a strange article appeared in the *Wall Street Journal* on September 14, 1989 [4]:

> Here is a space trip that's pretty far out even if it doesn't reach Neptune. The National Aeronautics and Space Administration will soon hunt for pyramids on Mars... The space agency says it has granted a longstanding request from some space buffs, who call themselves the Mars Project, to photograph the Cydonia region of Mars. Cydonia may be home to a field of pyramids and other rock formations erected by alien beings, speculate Mars Project enthusiasts, who pore over old photos of Mars looking for odd things. They have even found one rock formation that looks like a human face.

The article refers to Hoagland's ideas as "ruminations [that] have long been a source of merriment among space scientists." That a formal request from the head of the U.S. House of Representatives Science, Space, and Technology Committee to NASA would be reported in a reputable newspaper like the *Wall Street Journal* in this way was bizarre. Even more bizarre was NASA's response to it. According to NASA's chief Mars Observer project scientist, Arden Albee, "We'll try [to image the Face], if only to kill off the rumors."

So, as the launch of Mars Observer approached, NASA's true intentions concerning Cydonia were unclear. On one hand they had agreed, grudgingly, to re-photograph the area, but on the other, the same NASA scientists responsible for taking the pictures were stating publicly that they were sure the features in Cydonia were natural and not worthy of further study.

Mars Observer was finally launched from Cape Canaveral on September 25, 1992. For the next eleven months the spacecraft coasted toward Mars. Then on August 21, 1993, three days before entering orbit, contact was inexplicably lost.

While JPL was trying to re-establish contact with the spacecraft, NASA administrator Daniel Goldin ordered a formal review of the loss of the Mars Observer. After several months of deliberation, the review board, headed by Timothy Coffey, Director of Research at the Naval Research Laboratory, speculated that the most probable cause of the loss was a leak of pressured hydrazine fuel and helium gas, which caused the spacecraft to spin. As the spin rate increased, the spacecraft entered a 'contingency mode' in which the normal sequence of commands stored in the craft's on-board computer are suspended. The transmitter, which had been powered down by ground controllers, was never turned back on by the computer. But in reality, no one knew for sure:

> Because the telemetry transmitted from the Observer had been commanded off and subsequent efforts to locate or communicate with the spacecraft failed, the board was unable to find conclusive evidence pointing to a particular event that caused the loss of the Observer.

Adding to the mystery was the apparent lack of any real effort on the part of NASA to regain contact with the spacecraft. Hoagland devotes many pages in his book, *The Monument of Mars*, to the chronology of events surrounding this incident, including discussions within NASA about rebooting the on-board computer, using optical telescopes to detect bursts of infrared light from Mars Observer's laser altimeter, and turning on and listening for faint radio signals from another experiment, the Mars Balloon Relay. MBR was a self-contained 'black box' containing its own power supply, radio receiver, and transmitter. According to DiPietro,

> The urgent need to utilize this device as a beacon was apparent from the moment the spacecraft failed... It seemed logical that this device could be turned on and arrangements for detection could be made using a very large array antenna on Earth. This might have determined the whereabouts of the Mars Observer, in the same fashion as a survivor at sea... [5]

For some unknown reason, the decision to activate the MBR was not made until a full month after contact was lost with the spacecraft. But it was too late. As Mars and the Observer were approaching solar conjunction, with the Sun between them and Earth, any signal from the spacecraft would be drowned out by the overpowering radio noise of the sun. Even if Mars Observer was still intact and MBR still functioning, no one would be able to hear it. Subsequent attempts to contact the probe after it emerged from behind the sun were unsuccessful.

The McDaniel Investigation

The inexplicable loss of Mars Observer, together with NASA's strange behavior concerning the Face and other enigmatic features on Mars, led to talk of hidden agendas and government cover-ups. What began as a seemingly well-defined scientific question was turning into a highly-entangled and emotionally-charged political issue.

At first, the reason for this seemed to be the lack of new data. In a NASA fact sheet about the Face on Mars[14], it is stated that "...despite the phenomenal nature of such a potential discovery, no one in the scientific community... has ever proposed an investigation for a mission to study these features. Until more data are gathered, many scientists consider the probability that the features are anything other than natural in origin are just too low to justify the major expenditure of public funds which such an investigation would entail..."

But to get more data some public funds would have to be spent. This circular explanation reminded me of a conversation between a psychology professor and a post-graduate student:

Student: "There seems to be a need for good quality research in the field of hypnosis."

Professor: "I would never allow it in my department."

Student: "Why not?"

Professor: "Because hypnosis is not a respectable field for research."

Student: "Why not?"

Professor: "Because it has no serious published literature."

Student: "Why is there no literature?"

Professor: "Because nobody has done the research."

Student: "Why has nobody done the research?"

Professor: "Because it's not a respectable field of research."

In the fall of 1992, Stanley V. McDaniel, a professor of philosophy at Sonoma State University, became interested in the Cydonia investigation. McDaniel's specialty was epistemology, the branch of philosophy that investigates the nature and origin of knowledge. He remembered the Face from the Viking mission and wondered if NASA had any plans to re-image it with the Mars Observer. In his search for an answer, he first came upon Torun's analysis of the D&M Pyramid, Hoagland's book *The Monuments of Mars*, and later my book, *The Martian Enigmas*. But from NASA all he could find were statements to the effect that planetary scientists had determined the Face and other objects on

---

[14] http://venus.hq.nasa.gov/office/pao/facts/HTML/FS-016-HQ.html

Mars to be natural and not worthy of further study. Naturally he wondered if there were any scientific studies supporting their conclusions.

Even on the most basic question — had NASA examined these objects in detail? — McDaniel could not get a straight answer. When Mars researcher Dan Drasin asked Michael Malin about the Face and other anomalies in a letter written in 1992, Malin said, "The best scientific evaluation available today finds that there is no credible evidence to support the contention that these features are artificial." But another statement by Malin on one of his web pages seems contradictory, "No one in the planetary science community (at least to my knowledge) would waste their time doing a 'scientific study' of the nature advocated by those who believe the 'Face on Mars' artificial."

Several of us had published the results of our studies in peer-reviewed professional journals, yet NASA seemed completely unaware of our work. In response to a 1993 letter from Senator John Glenn on behalf of a constituent interested in the features, NASA wrote,

> The resemblance to a human face is due to the particular lighting angle at which the images were taken. This conclusion is supported by the fact that the 'face' disappears in images of the same place taken at different lighting angles.

Even after DiPietro and Molenaar had found the second confirming photograph of the Face taken at a different lighting angle, and I had shown, using shape-from-shading, that the Face was not an optical illusion, NASA continued to say it was.

It became clear to McDaniel that no scientific study had ever been performed by NASA. The only document he could find was an anonymous NASA memorandum entitled "Technical Review of the Monuments of Mars" which had been sent to Mars activist David Laverty from Mark A. Pine, Chief of the Policy and Plans Branch within NASA's Office of Space Science and Applications. According to McDaniel,

> This memorandum cannot be taken seriously as a responsible scientific evaluation. It refers only to a limited selection of claims made in a single work on the subject (a popular book not intended as a strict scientific report). The claims that are dealt with are taken in isolation, generally misrepresented, and their evaluations are cursory and significantly flawed. Although the paper is characterized as a technical review, it does not deserve the title by any responsible standard. The use of it in an official communication sent out by NASA in response to an inquiry by a United States Congressman raises a very serious concern about the integrity of NASA's treatment of the subject. The question of integrity is compounded by NASA's distributing apparently false claims about photographs to members of Congress and the public, and by its apparent resort to ridicule in place of coherent scientific evaluation [6].

During the course of McDaniel's investigation, in an attempt to improve communication and reduce some of the apparent confusion over Cydonia, several of us contacted Malin directly. In the same letter mentioned above, Dan Drasin asked: "To what extent have you personally examined and evaluated the specific observations and claims regarding the Cydonia anomalies, and if so, what are your specific reasons for rejecting them as potentially evidential?" About the Face, Malin says this:

> Investigations of the 'Face' originally cited its 'bilateral symmetry' as proof of its artificial nature. However the landform is clearly not bilaterally symmetric. One need only hold up a pocket mirror to the figure on page 21 of Dr. Carlotto's book [first edition of The Martian Enigmas], and compare the form of the feature as reflected in the mirror with the actual side of the 'face', to see this. Once this was pointed out, the proponents of the artificial hypothesis constructed plausible reasons for the lack of symmetry. However, since the 'observations' failed what was originally believed to be a substantial test, and the explanations for this failure are ad hoc, it remains simply people's opinion that it looks like a face. I have rocks that look like faces, but they were not fabricated — they simply eroded to look like faces. The figure on page 41 of Dr. Carlotto's book is, I think, excellent evidence that the Face is simply a funny looking hill — there is nothing unusual about it.

Bilateral symmetry was never offered as the only proof of artificiality. As we saw in a previous chapter, the Face has other unusual qualities: humanoid-like proportions, placement on a highly symmetrical platform, and fine-scale detail — detail unlike that seen in any other mesa or knob in the area. Malin did not comment on any of these qualities, or on the results of the shape-from-shading experiment that showed the impression of a face persisted over a wide range of lighting conditions and viewing geometries. Malin did, however, mention the fractal analysis, which showed the Face to be the least natural object in the area:

> Fractal measurements basically determine the difference in 'texture' of a surface relative to some model of the 'natural background'. For the Viking images, Dr. Carlotto used as 'background' either an average of each image, or the smooth areas that occupy most of each image. It doesn't take a computer to see that the mesas and buttes are texturally different from the average. Further, the fractal analysis does not demonstrate something is artificial, it demonstrates that it is different. I suggested to Dr. Carlotto that he apply his technique to several other areas of Mars, some of which have no 'anomalous' features, to see if he gets alarms. I suspect that he will (again, on things that differ from the average).

Based on these comments, it was clear that Malin did not understand the fractal technique. According to McDaniel, "Given the clear presentation of the technique in Dr. Carlotto's book, his misunderstanding is puzzling." As mentioned in an earlier chapter, after telling Carl Sagan about our work with fractals, he sent me a copy of a paper he had written twenty years earlier on a similar idea — that deviations from thermodynamic equilibrium could be used as a possible indicator of intelligence. Just as Sagan noted that "thermodynamic disequilibrium is a necessary but of course not a sufficient condition for the recognition of extraterrestrial intelligence" [7] we were not claiming the Face's non-fractal behavior to be definitive proof of its artificiality, but only another piece of circumstantial evidence.

Shortly after my book, *The Martian Enigmas*, was published, I sent Malin a copy of it along with a reprint of the fractal paper Mike Stein and I had published in the *Journal of the British Interplanetary Society* a few years earlier. During the review process for the paper, one of the referees wrote that "to make a convincing case for the Cydonia objects it is necessary to show that the technique does not predict a plethora of non-natural objects in the Viking images." In order words, we had to show the technique did not generate 'false alarms'. We extended our analysis to several additional Viking frames, an area in excess of 15,000 square kilometers, and found the Face was still the least fractal object by a considerable margin. That the paper was ultimately accepted and published by JBIS would seem to indicate that the area processed was sufficient to convince at least an impartial reviewer of the significance of our results.

McDaniel's assessment of the situation with NASA was not flattering [6]:

> As my study of the work done by the independent investigators and NASA's response to their research continued, I became aware not only of the relatively high quality of the independent research, but also of glaring mistakes in the arguments used by NASA to reject this research. With each new NASA document I encountered, I became more and more appalled by the impossibly bad quality of the reasoning used. It grew more and more difficult to believe that educated scientists could engage in such faulty reasoning unless they were following some sort of hidden agenda aimed at suppressing the true nature of the data.

How Alien is Alien?

Concerning planetary exploration in general, Malin had said that "significant discoveries remained to be made by cameras." Yet in his reply to a question by Drasin concerning the use of his Mars Orbital Camera in attempting to shed further light on the specific question of the Martian anomalies Malin seems to contradict himself:

> ...what will Mars Observer be able to do to resolve the differences of interpretation? Unfortunately, I am not as sure it'll help as are Mr. Hoagland and Dr. Carlotto. Even if we get a picture of the 'D&M Pyramid' or the 'Face', what will such an image prove? ... I hope people will state what criteria would convince them that they are wrong (as well as what criteria would convince them that they were right) before we get to Mars. I don't think confirming previous angle and distance measurements will constitute validation of their hypotheses; one must find something substantially different or unanticipated. Let me pose two scenarios:
>
> 1. Mars Observer Finds roads and buildings
>
> 2. Mars Observer finds rocks, cliffs, channels, and dirt.

In Case 1, I believe the results are unambiguous. Mr. Hoagland, Dr. Carlotto, and their associates can rightfully claim their accolades and laurels. What, however, happens in Case 2? Will they admit they were in error? Will they abandon their work? I think not.

Malin's goes even further in an August 1994 article in *Omni* magazine [8], in which he states that a high-resolution photograph near the face showing "roads or large areas that have been excavated will prove his hypothesis wrong. On the other hand, if we see just a natural-looking surface, then I would argue my hypothesis is correct." Like Sagan and Wallace in their 1971 paper on the recognition of terrestrial intelligence in satellite images, Malin expected to find artifacts of recent habitation — terrestrial features such as roads and buildings.

We are the only intelligent life form we know of; therefore, it is difficult to extrapolate from our limited experience to predict what alien artifacts would look like. So it seems a safe bet to expect the unexpected, to be on the lookout for anomalies — surface features that stand out from the background in terms of their geometrical structure (linearity, rectilinearity, or parallelism), symmetry, periodicity (regular repetition of a form), surface smoothness, texture, or other characteristics. Should artificial structures on Mars exist, the challenge of detecting them is compounded by the likelihood that they are probably ancient and highly eroded, perhaps to the point of being barely distinguishable from the background geology. It is likely that any indication of artificiality would be very subtle. This, of course, leads to the question of how can we distinguish highly eroded archaeology from geology — a question that will be considered later in the book.

The Ethics of Scientific Debate

Late in 1995, I established a web page on the Face and other objects in Cydonia. It contained information on shape-from-shading, image enhancements, plus a bibliography of all of the research papers and books that had been written on the subject. Around the same time, MSSS established their own Face on Mars web page. The only reference it contained was a link to JPL's original 1976 press release concerning the Face. Like Sagan's *Parade* article "The Man in the Moon," the MSSS web site contained no bibliography and made no mention of any of our work, other than to say "some people have argued, mostly in the lay literature, that the face-like hill is artificially shaped"[15].

But of greater concern were a series of computer generated views of the Face rendered, as I had done, under different lighting and viewing conditions. The MSSS web page never commented on my shape-from-shading analysis of the Face published in *Applied Optics*. Instead, it gave the public its own version, based on an "admittedly lower spatial resolution" reconstruction of the shape of

---

[15] http://barsoom.msss.com/education/facepage/face.html

the Face. The resolution was so low that the Face did not look like a face even when it was rendered under the original lighting conditions.

Recall that computer graphics techniques generate images from 3-D models. Shape-from-shading is inverse computer graphics, computing a 3-D model of the shape of an object from its 2-D image. The most basic test of a shape-from-shading algorithm is whether it is able to accurately regenerate the original image from the computed shape. A more sophisticated test is to predict what another view (e.g., a different lighting condition) would look like. This is what O'Leary and I had done in 1988 to validate our results. That the MSSS surface reconstruction could not even pass the most basic test calls into question the accuracy of the visualizations on their web page — visualizations, which, because of their lack of detail, show nothing unusual about the Face.

An even more blatant misrepresentation of our findings was discovered by McDaniel on another MSSS web page that discussed the processing of Viking orbiter images. Malin presents an example taken from a popular supermarket tabloid on the "pitfalls of image processing." The magazine article mentions the 'teeth' that I had discovered in 1985. Malin begins by saying that one cannot extract more information from an image than there is in the beginning. He then goes on to show how the Face got its teeth. By an excessive amount of contrast enhancement he is able to make pixel noise in Viking frame 70A13 look like teeth.

To the uniformed reader it seems to make sense. But there is a problem — Malin's teeth are false. The ones I found were on the other side of the Face. Moreover, they are present in enhancements of both 35A72 and 70A13 images. Their presence in both images, like other fine-scale detail in the Face, suggests they are real features and not noise.

McDaniel wrote Malin three times in an attempt to get him to correct the many errors on this web page, but to no avail. In a web-based article, "Dr. Malin's False Teeth," written a year later McDaniel states:

> The ethics of scientific debate indicate, first of all, that one should identify those researchers with whom one takes issue. Dr. Malin does not name either Dr. Carlotto or DiPietro and Molenaar in his discussion of the 'teeth'... What Dr. Malin has done, whether intentionally or not, is to set up what is called a "straw man" argument — a fallacious argument that addresses a misrepresentation of the view under discussion rather than the actual view.
>
> Second, it is a scientific obligation to get the facts straight, and make the appropriate references to the literature... If he were to make these proper references to the literature, Dr. Malin would certainly never present his 'false teeth' as though they are the actual features referred to when the 'teeth' in the Face are under discussion. He also would never be inclined to present enhancement errors as the procedure used by these expert professionals...

Certainly the public must be disturbed when the Principal Investigator for the Mars Global Surveyor Camera, who bears a heavy responsibility as the single individual upon whom the decision to obtain new high resolution images of the Cydonia features rests, presents himself to the international public as being apparently misinformed, as taking his data from supermarket tabloids, and as seeming to participate in a long-standing pattern of apparent misrepresentations by NASA...

The Mars Orbiter Camera

In contrast to his flawed and superficial understanding of our work, Malin went into considerable detail on the many reasons why it will be difficult for his camera to image an isolated object such as the Face. Seven reasons were cited in an email[16]:

1. The MOC is body-fixed to the spacecraft

2. The MOC has a limited cross-track field of view

3. The MOC has a large but not "infinite" along-track field of view

4. The spacecraft has limited pointing control

5. There will be a substantial uncertainty in the predicted inertial position of the spacecraft (and hence, the camera)

6. The non-inertial position of the spacecraft will also be uncertain

7. The spacing of orbits will be uncertain

The MOC really contains two cameras: a high-resolution narrow-angle camera capable of imaging objects as small as 1.4 meters across from an altitude of 400 km, and a low-resolution, wide-angle, two-color camera with a ground resolution of about 280 meters per pixel. Each camera uses a linear photodetector array which forms a picture by scanning the ground one line at a time, like a photocopier. The size of a scan line on the ground (the width of the picture) depends on the camera's cross-track field of view — about 3 km at an altitude of 400 km for the narrow angle camera. The length of the picture is limited by the amount of on-board memory available to store all of the scan lines which make up a picture. Malin gives a length of about 15 km on the ground for the narrow angle camera under normal conditions.

Because the MOC is bolted to the spacecraft and has no independent pointing capability, in order to aim the camera at a particular target, the attitude (i.e., pitch, yaw, and roll) of the spacecraft must be changed. The spacecraft itself has a pointing accuracy of about 0.6 degrees, which translates into a 4 km uncertainty on the ground. Orbital uncertainties of 40-120 seconds in the time the spacecraft will be at a specific point in its orbit lead to 120-360 km

---

[16] http://www.skepticfiles.org/skep2/marsobs1.htm

positional uncertainties along the orbital track, and a 7.4 km cross-track uncertainty at 40° N latitude under the best conditions. Also, because the position of Mars' longitude/latitude grid is also uncertain, the location of specific features on the ground cannot be determined to better than about 5-10 km. Finally, the exact spacing of the orbits will be uncertain. At 40 degrees latitude, where the nominal spacing is roughly 2.4 km, during the course of the mission a given target would be imaged, at most, twice by the narrow-angle camera.

He argues that because the magnitude of these uncertainties exceed the MOC's narrow-angle camera's field of view, "hitting anything as small as a specific 3 km piece of the planet is going to be very difficult," but adds that the

> MOC team is attempting to address some of these issues with, for example, optical navigation. This could reduce the spacecraft position uncertainty by perhaps a factor of five or more. An attempt will be made to generate a new control grid with higher precision (perhaps as good as 1 km). But nothing can be done about the orbit spacing or the pointing control or the width of the MOC field of view.

Mars Global Surveyor

Following the loss of the Mars Observer in the fall of 1993, NASA scrambled and, in just a few months, developed plans for a follow-up mission to Mars, the Mars Global Surveyor (MGS). Using spare parts from Mars Observer, work started on MGS in October 1994. The spacecraft contained six of the eight original Mars Observer experiments [9]. They included the MOC, Mars Observer Laser Altimeter, Thermal Emission Spectrometer, Magnetometer/ Electron Reflectometer, Ultra Stable Oscillator (for radio science), and Mars Relay system. To reduce costs, MGS had to be down-sized so that it could be launched aboard a smaller Delta rocket. And to save weight, the spacecraft would 'aerobrake' on reaching Mars by using the Martian atmosphere to slow it down and gradually place it into the proper orbit.

MGS was launched on November 17, 1996. The plan was for it to be ready to support Russia's Mars 96 mission, which was to be launched a day earlier. Unfortunately, the fourth stage of the Russian's Proton booster failed. The Mars 96 spacecraft re-entered Earth's atmosphere and crashed into the Pacific Ocean off the coast of Chile the next day.

MGS reached Mars on September 11, 1997. At 6:31 Pacific Daylight Time, the main rocket engine fired for 22 minutes, slowing the spacecraft to place it into a highly elliptical orbit around Mars. On September 16, another burn occurred, moving the low point of the orbit from 250 km down into the Martian atmosphere to begin aerobraking. Aerobraking continued for the next 24 days, reducing the orbital period from 45 to 35 hours, but was temporarily suspended on October 11 when flight controllers decided to raise the orbit to study a possible problem that had developed with one of the solar panels. After a two week hiatus, aerobraking was resumed on November 7, but at a more gradual rate in order to reduce risk of damage to the solar panel.

Since the start of aerobraking in September, Malin's MOC had been photographing the Martian surface. By early November well over a hundred pictures had been taken, yet only six (about 5%) were released. This trickle of MGS imagery was in stark contrast to the steady stream of Pathfinder lander images that had been posted on the Internet the previous summer. The lack of imagery was particularly troubling in light of the Cydonia controversy. MSSS and JPL indicated that it would be difficult to image isolated features such as the Face; however, an analysis of the few images that were released suggested the camera was a much more capable instrument than we had been led to believe.

On his web page Malin said that "The MOC is body-fixed to the spacecraft," and "The spacecraft has limited pointing control." This implied that it would be difficult to aim the MOC at a specific target. What he did not say was that the attitude of MGS is controlled using a set of three reaction wheels, not thrusters. Changing the spin of a reaction wheel changes its angular momentum, which in turn, causes the spacecraft to rotate around the corresponding axis. Even though the MOC cannot be positioned independently of the spacecraft, the spacecraft itself can be moved with great ease and flexibility without using any fuel. That MGS was being maneuvered a considerable amount was evident in the first four targets imaged by the MOC: Labyrinthus Noctis, Nirgal Vallis, Valles Marineris, and Olympus Mons, targets where the camera had to be tilted up to 35 degrees off the orbital track.

Malin stated on his web page that MSSS was exploring ways of reducing the large targeting uncertainties. Because a high-resolution image-map of Mars did not yet exist, MSSS had begun in 1994 to register Viking imagery to the USGS Mars Mosaicked Digital Image Model[17]. The intent was to use these map-projected Viking images for target planning.

In each of the four MOC images released by MSSS, its corresponding image 'footprint' was shown within a lower resolution, map-projected Viking orbiter 'context' image. Because of the narrow field of view of the high-resolution camera, context images provided a visual reference for interpretation. The close match between the MOC image footprints in the context images, which were the same map-projected images used for target planning, and the actual MOC image that was collected indicated MSSS had reduced their originally stated 12.4 to 17.4 km targeting uncertainty by a considerable amount.

I wondered whether Cydonia could have been targeted during the initial aerobraking phase from September 15 (orbit 3) to the suspension of aerobraking on October 6 (orbit 15). Aerobraking occurs at periapsis (closest approach) when the spacecraft dips into, and is slowed down by, the

---

[17] http://www.msss.com/mars/observer/camera/papers/gds_papers/geodesy/geonav.html

atmosphere. According to MSSS, following periapsis, a 'roll-out' maneuver from aerobraking to 'array-spin normal' orientation occurs. It is only at this point in the orbit, shortly after periapsis, that the camera can be used. Since images could only be acquired just after periapsis, imaging opportunities would have been severely limited. It seemed unlikely that MGS would have had the opportunity to image Cydonia during this initial aerobraking phase because Cydonia was probably too far to the north.

However, during the aerobraking hiatus there was much greater flexibility in changing the attitude of the spacecraft and more opportunities to aim the camera at Cydonia. The orbital period during the hiatus was about 35 hours. Given the rotational period of Mars (about 24 hours), MGS's orbit precessed 154 degrees in longitude to the east on each revolution. If the first orbit passed over the equator at some longitude, say zero degrees, the longitudes for the next seven orbits would be 154, 308, 102, 256, 50, 204, and 358 degrees. Putting these in order, we get 0, 50, 102, 154, 204, 256, 308, and 358 degrees. This rough 'back of the envelope' calculation showed that during the aerobraking hiatus, MGS passed within 52 degrees of longitude of any point on Mars. Since the operation of the camera was not limited to the roll out maneuver, it could have imaged just about any point on the surface to within the accuracy of the targeting software.

MGS was clearly being tasked by MSSS to point the MOC toward a predetermined set of targets. That Cydonia could be one of these targets seemed a distinct possibility.

Targeted Observations

Following publication of the McDaniel Report in 1993, Stanley McDaniel organized the Society for Planetary SETI Research (SPSR), a group of about two dozen scientists and other professionals from a wide range of disciplines devoted to the search, analysis, and evaluation of possible evidence of extraterrestrial artifacts within our solar system, with the near-term focus being Mars, and the array of objects in Cydonia.

On November 24, six representatives from SPSR: John Brandenburg, Horace Crater, Vincent DiPietro, Stanley McDaniel, David Webb, and I, met with Carl Pilcher, the Acting Director of Solar System Studies, and NASA scientist Joseph Boyce. The hour-long meeting, which had been arranged by Brandenburg and McDaniel achieved its goal. In Webb's words:

> Pilcher firmly and unequivocally stated that it was the official policy of his department and of NASA that the Cydonia region was to be imaged at high resolution during fly-over and that Glenn Cunningham and Mike Malin were aware of, and had signed off on the policy... He went on to say that everyone was interested in having the area imaged: one group because they wanted to show us how wrong we are and have been all along; the other group, because they feel that

we have some interesting material, and they would like to see just how interesting it turns out to be.

As aerobraking continued, a study of the solar panel problem indicated the damage was more extensive than originally thought. In the meantime, a new plan began to take shape. The ultimate goal of aerobraking was to gradually place MGS into a sun-synchronous mapping orbit, a near-circular two-hour orbit designed so that the spacecraft would pass over the equator at the same local time, about 2 PM. In order to reduce further damage to the solar panel, the decision was made to continue aerobraking at the reduced rate until late March or early April, and then leave MGS in an interim 'science phasing' orbit until the fall. A second phase of aerobraking would commence then and last for another three months. At the end of this phase MGS would be in an alternative mapping orbit, with the spacecraft passing over the equator twelve hours later, at 2 AM local time instead.

On March 26, McDaniel received a telephone call from MGS Project Manager Glenn Cunningham informing him that the spacecraft would come out of aerobraking that evening and enter the science phasing orbit. While in this orbit, MGS would perform targeted observations of selected features on Mars[18]. During the following month there would be three opportunities to image the objects in Cydonia: on April 5, 14, and 23 UTC (universal time)[19]. The Face was at the top of the list.

A few days later, McDaniel and I met with Cunningham and Arden Albee at JPL to discuss their plans for imaging and data release. We learned that the pictures would be taken in the morning, with the sun higher in the sky than it was in the Viking images. We voiced our concern that this could reduce contrast and shadowing, making the imagery harder to interpret. But assuming a good picture of the Face, or another one of the features of interest, was obtained, we were curious about how JPL would handle three possible outcomes: 1) obviously artificial, 2) obviously natural, or 3) not sure. They said the plan was to simply release the data and say nothing, letting the picture speak for itself. JPL would stay out of the controversy.

Several days after our meeting at JPL, I generated a prediction of how the Face might appear based on an estimate of the MGS imaging geometry, provided to me by Mars researcher Peter Nerbun. In 1988, as part of my shape-from-shading analysis, I computed synthetic images of how the Face appears under different lighting conditions. The images showed that under certain conditions, particularly those with the sun high in the sky (late morning to early afternoon), the impression of a face was not very distinct. According to Nerbun's model, the sun would be illuminating the face from below. As it turned out, this was just about the worst of all possible lighting geometries.

---

[18] http://mars.jpl.nasa.gov/mgs/target/cydoniapress.html
[19] http://mars.jpl.nasa.gov/mgs/target/announcement2.html

Ironically, it was only because of the problem with the spacecraft's solar panel that the opportunity to photograph Cydonia so early in the mission had become available. After waiting for more than twenty years for new images of Cydonia, that the lighting conditions were not going to be ideal seemed relatively unimportant at the time.

With the April 5 target date approaching, McDaniel warned on his web site,

> We have frequently stressed that it is not the position of SPSR scientists to advocate any particular outcome, but to go with the facts as they emerge. If the objects turn out to be clearly natural in origin, we will be among the first to say so. However, it is also important to avoid premature judgment based on initial impressions. It may take many months of study, augmented by data coming from the mapping mission later in 1999, before the true impact of the MGS mission on the question of the anomalies will be understood.

For better or worse, a lot was riding on these new pictures.

Sources

1. Thomas S. Kuhn, *The Structure of Scientific Revolutions*, The University of Chicago Press, 1970.
2. M.C. Malin, G. E. Danielson, A.P. Ingersoll, H. Masursky, J. Veverka, M.A. Ravine, and T.A. Soulanville, "The Mars Observer Camera," *Journal of Geophysical Research*, Vol. 97, No. E5, May 25, 1992.
3. M. Caplinger, "The Mars Observer Camera Ground Data System," *9th AIAA Conference on Computers in Aerospace*, October 1993.
4. R. Davis, "One thing we know right now is just who's going to pay for it," *Wall Street Journal*, Sept. 14, 1989.
5. *The Case for the Face, Scientists Examine the Evidence for Alien Artifacts on Mars*, Stanley V. McDaniel and Monica Rix Paxson, Editors, Adventures Unlimited Press, Kempton IL, 1998.
6. S. V. McDaniel, *The McDaniel Report*, North Atlantic Books, Berkeley CA, 1994.
7. C. Sagan, "The recognition of extraterrestrial intelligence," *Proceedings of the Royal Society*, Vol. 189, pp 143-153, 1975.
8. R. Kiviat, "Casting a New Light on the Mars Face," *Omni*, August 1994.
9. Fred G. Komro and Frank N. Hujber, "Mars Observer Instrument Complement," *Journal of Spacecraft and Rockets*, Vol. 28, No. 5, September-October, 1991.

Eight — New Evidence of Artificiality in Cydonia

> There is a sharp disagreement among competent men as to what can be proved and what cannot be proved, as well as an irreconcilable divergence of opinion as to what is sense and what is nonsense." — Eric Temple Bell

In the Pacific time zone, it is just after midnight on April 6, 1998. Mars Global Surveyor has just passed over Mars' north pole and is heading south. Just before reaching periapsis — the point of closest approach to the surface — the spacecraft rotates slightly toward Cydonia and the Face. At 12:39 AM, the camera is switched on. As the craft flies over the Martian surface, an image 4.5 km wide by 42 km long is formed and stored in an onboard computer.

Most of the time, I 'telecommute' by way of the Internet between my home, north of Boston, and my corporate office in northern Virginia. But today on April 6, I am up at dawn to fly to Washington DC for a series of meetings over the next couple of days. By the time I arrive at my office mid-morning, MGS has already begun to transmit the picture of Cydonia by way of NASA's Deep Space Network to JPL.

Between meetings, I nervously check the MGS web site. Just after lunch, the raw unprocessed image over Cydonia is posted[20]. Heart racing, I begin to download the data. After getting most of it, about 9 megabytes, my computer runs out of memory. Freeing up some space, I try again, and finally get the image. By this time, several others in the office have gathered around. As the image comes up on the screen, all we see is a pattern of vertical stripes. Being a raw, unprocessed image, I realize the stripes are caused by gain variations across the camera's photodetector array, and quickly write a destriping program. By mid-afternoon the image has been cleaned up and contrast stretched.

A colleague of mine casually walks by and asks, "Where's the Face?" Because of the extreme viewing geometry and the unusual direction of illumination, it is not immediately apparent. But as I stared at the image, I realized my prediction from the preceding week did match its general appearance. Being lit from below, the effect was like holding a flashlight under your chin. The impression was completely different from the Viking images. Where the Viking images were taken almost directly overhead late in the afternoon during the Martian summer, the new MGS image was taken mid-morning during the Martian winter with the spacecraft viewing the western side of the Face from a point 45 degrees above the horizon (Table 3). The new image was significantly different in appearance (Figure 41). It was obvious that the Face, if it was in fact intended

---

[20] http://mars.jpl.nasa.gov/mgs/target/CYD1/index.html

to represent a face, was severely eroded. Maybe it was a geological formation after all. Still, there was much about it that was unusual.

When I got back to my hotel room that night, there was a message from John Brandenburg and Vince DiPietro, "Call us. We are puzzled and going to get drunk."

Figure 41 Portion of MGS image 22003 (left) and corresponding Viking image 35A72 (right) containing the Face and the northeast quadrant of the D&M Pyramid The images are hard to correlate because of the obliquity of the MGS image, and the presence of thick haze that day over Cydonia. (JPL/MSSS)

Table 3 Acquisition geometry for the Viking and MGS images. These two images could not be more different in terms of season, time of day, imaging geometry, resolution, and atmospheric state.

Parameter	MGS (22003)	Viking 35A72
Aerocentric Solar Latitude (Season)	303.3° (Winter)	99° (Summer)
Incidence angle (Solar elevation)	25°	80°
Solar azimuth	159°	268°
Emission angle (Spacecraft elevation)	45°	11°
Spacecraft azimuth	232°	169°
Resolution	4.3 m	47 m

The Face Disappears

The week before, McDaniel had posted a report on his web site concerning our meeting at JPL with Cunningham and Albee. At the meeting, we learned that the sun would be higher in the sky than it was in the 1976 Viking images, and illuminating the Face from below. McDaniel voiced our concern:

> If the sun is too high, this could obliterate shadows and make the facial appearance indistinct. Should this be the case, we would not expect skeptics to claim this shows 'there is no face there', since it is well-established already by means of synthetically generated views that at certain sun angles the features of a sculptured face could be difficult to see.

We should have known better.

A few hours after the MGS image was received, Timothy Parker at JPL's Mission Image Processing Laboratory posted a 'contrast-enhanced' version of the Face. Although image enhancement is supposed to improve the visual quality of a picture, JPL's enhancement made it look worse — the Face looked like scratches in the sand. This image, which Mars researcher Lan Fleming calls the 'catbox' image, appeared on the news that night (Figure 42).

As promised, JPL made no official comment. They didn't have to. Their picture said it all. No one, not even I, could see a face in their 'enhancement'. Was it simply a bad job, done in haste to get a quick picture out to the public and the media, or a clever move by certain individuals to put the matter to rest once and for all?

We all wanted the matter to be resolved, but not by the media in a few hours on the evening news. McDaniel seemed to see it coming. Again, from a posting on his web site the week before,

> It is my personal view that the problematic way NASA has treated this subject in the past is the result of certain unfortunate moves made early on in the history of this debate, again because of premature conclusions and only cursory analysis of the data. We urge NASA scientists, as well as members of the public who may attempt analysis and interpretation of the data, to view the data objectively and with care, avoiding premature conclusions. The stakes are too high to allow bias or a desire to 'be first' to obscure the truth that may eventually emerge from the data provided by the Surveyor.

Clearly over the years, planetary scientists had become irritated by the Cydonia controversy — a debate they felt had gone on for too long. Shortly after the MGS image of the Face was released, geologist Michael Carr expressed their sentiment, "I hope we've scotched this thing for good."

It seemed as if they had.

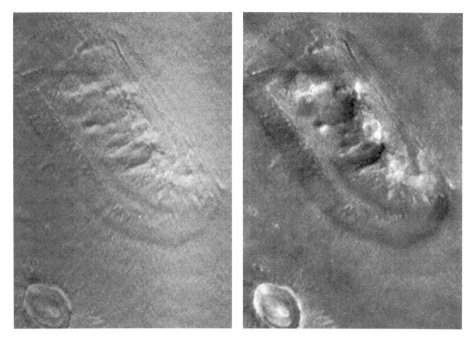

Figure 42 JPL's 'catbox' image (left) resulting from excessive contrast enhancement. More accurately restored image (right) with slight contrast stretch to emphasize shading.

Restoring the Face

I spent a considerable amount of time trying to understand this new picture of the Face. After cleaning up the image, the surface appeared obscured in places by thin clouds and haze. Unlike the Viking images, which were acquired in summer, the new MGS image was taken in late winter. At this time of year, haze, clouds and dust in the atmosphere are not uncommon in Mars' northern hemisphere. In fact, a few days earlier, on April 3, MGS had attempted to photograph the location of one of the Viking landers, but found Utopia Planitia to be under a heavy overcast. Likewise, on April 5, clouds and haze were present over much of Cydonia, but by an amazing stroke of luck, there was a break over the Face.

Even though the Face was visible, it was illuminated mostly by ambient light scattered by the hazy winter atmosphere. Under these conditions, shadows are weak and visual acuity poor. Close examination of the image indicated the presence of bright areas that did not appear to be bright because of shading (i.e., because they were facing the sun). Instead, they seemed to consist of a more reflective material, perhaps frost[21]. The perception of form in a shaded image of

---

[21] The Viking landers found that patches of frost, most likely water ice, often form overnight and sometimes persist for extended periods in winter.

a 3-D surface degrades when the lighting becomes diffuse and when the reflectance of the surface varies. Together, the presence of atmospheric haze and variations in surface brightness due to the frost made the raw image very difficult to interpret.

One way to improve the appearance of an image is by a kind of image enhancement known as 'high-pass filtering'. High-pass filtering emphasizes higher frequency detail (texture), while suppressing the lower frequency (tonal) background. The cutoff frequency is critical — if it is too high, important grayscale information will be lost, making it difficult to perceive the 3-D shape of an object. In Parker's contrast enhancement, the low-frequency cutoff was too high and the image was completely washed out. The Face appeared flat and featureless. But by lowering the cutoff frequency, I found that most of the confusing variation caused by haze and frost could be removed while still preserving enough of the shading information to see the Face for what it was — an object with significant 3-D relief.

In JPL's enhancement, the Face disappeared. In my enhancement, it was still there, albeit eroded. Which enhancement (Figure 42) was more accurate?

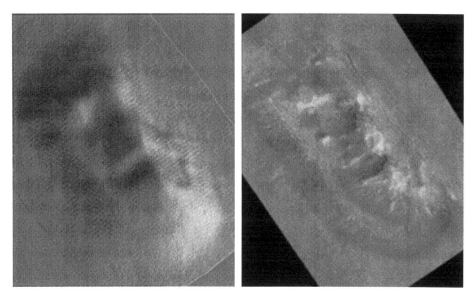

Figure 43 Predicted MGS shaded view rendered from Viking elevation model (left) and restored MGS (right).

A week earlier, I had generated a prediction of what we expected MGS to see on April 5 based on a 3-D model of the Face derived from the Viking data. I revised my prediction slightly, taking into account the actual MGS camera geometry and lighting conditions. The predicted image was produced by computing a shaded rendition of the Viking-derived elevation model using the sun angle at the time the MGS image was taken, and reprojecting it based on the location of the spacecraft. Although the Viking elevation model is about ten times lower in resolution and thus unable to predict fine-scale detail, that the prediction was more like my restored MGS image in general appearance (Figure 43) suggested my enhancement was a more accurate representation of the Face than JPL's.

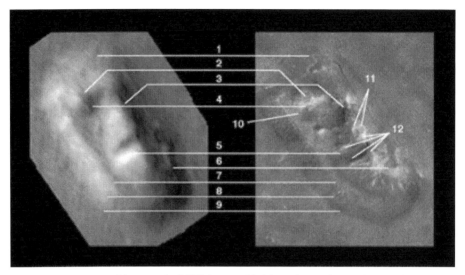

Figure 44 Features of interest in MGS image (right) and reprojected image from 70A13 (left). Features 1-9 are present in both Viking and MGS images, 10-12 are new features found in the MGS image.

Comparing Facial Features

Because the Viking and MGS images were acquired from different positions in space, it is necessary to reproject one image to match the other in order to compare them. JPL had never done this. After reprojecting the 70A13 image of the Face to match the MGS image a number of points of correspondence became evident (Figure 44).

1) Linear features on the crown of the head — These form a symmetrical 'X' pattern in the Viking images. In MGS, the bottom of the 'X' appears to be the top edge of the platform. The two upper arms of the 'X' are different. The left is straight while the right has several curves. Both appear to be cuts into the side of the formation leading from the platform down to ground level.

2) Eye brow — There is a sharp cliff just above the left eye region, which is bright in the MGS image because it is illuminated from below. In the Viking image, where it is lit from the upper left, it casts a shadow that produces the impression of the left eye. Because the Face is lit from below, this effect is missing in the MGS image. Whether the dark circular region discussed below is what remains of the physical eye structure is an open question.

3) **Nose ridge** — Evident in both images, it appears as a round sculpted feature in Viking. Narrow ridges on the left side of the ridge are evident in the higher resolution MGS image. A pattern seen as broad parallel stripes in the Viking images extends across the Face and correlates with these features in the MGS image.

4) **Pupil** — Within the dark region in the left (west) eye, a darker circular area can be seen in the Viking images, which looks like a pupil. In Viking image 70A11 there is a small bright structure to the left that appears to be casting a pupil-like shadow. This same object is seen as a bright region in the MGS image and appears to be a protruding object at the base of the cliff just above it. Whether the pupil is an illusion created by a shadow or an actual circular feature is unclear at this time.

5) **Mouth** — This feature is evident in both images. New lip-like features can be seen in the MGS image and are discussed below. Structures that looked like teeth in enhanced Viking images are not evident in the new MGS image. It is possible that they may not be visible when illuminated from below as suggested by Mars researcher Ananda Sirisena, or may not be present at all.

6) **Erosional features** — In both Viking and MGS images, a crescent-shaped structure can be seen on the right side of the Face. It runs from just below the chin up and over to the near the right side of the platform, where it appears to lead into a dark channel-like feature. The crescent shape suggests a dune, formed perhaps from windblown material deposited on the leeward side of the Face.

7) **Platform** — In both MGS and Viking images, the Face appears to lie on a relatively flat surface with a well-defined edge on the left side. MGS reveals the top edge of the platform to have a slight peak near the middle, like a crest, which corresponds to the bottom arms of the "X" pattern noted above (Figure 45). Because of shadowing in the Viking images, and the oblique view of the MGS image, the right side of the platform is not visible.

8) **Beveled edge** — This feature is a sloped transition from the flat platform to the ground — about 100 meters below. This type of edge is not unique in Cydonia — other landforms also posses the same general feature; however, the edge around the left side of the Face is much more regular than the edges of the other landforms. It is somewhat wider at the top of the head suggesting a more gradual slope from the crown of the head down to ground level.

9) **Base of platform** — In the Viking image the base of the Face appears to be very symmetrical. This is confirmed in orthorectified MGS imagery and measurements of lateral symmetry discussed later.

Besides these features common to both the Viking and MGS images, several new features are visible in the MGS image:

10) **Dark circular region** — In the MGS image, there is a dark region just below the left (west) pupil that cannot be seen in the Viking images. It could be either a slight depression, which appears darker because it is sloped away from the sun, or a region containing somewhat darker material. That it is located in the general area of the left eye begs the question whether it could be the remains of the physical structure of the left eye. The corresponding area on the right side of the Face has not been directly observed by either Viking or MGS.

11) **Nostrils** — Corresponding to a flattened area near the tip of the nose ridge in the Viking images is a pair of circular depressions that look like nostrils. These features appear to be on either side of lateral axis of symmetry of the Face (Figure 45).

12) **Lip structures** — Rather than being a sharp cut through the central ridge, the top and bottom of the mouth slope gradually down to the level of the platform. The shape of these sloped areas are not unlike lips. Adding to this impression is what could be referred to as a "harelip" which comes together between the base of the nose and the mouth.

After taking the time to carefully examine the new MGS image, it became clear that it not only confirmed the facial features seen in the older Viking images but revealed additional details consistent with the hypothesis that the Face might be an artificially-constructed representation of a humanoid head.

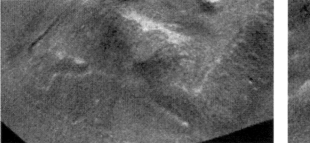 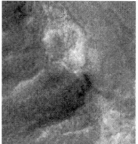

Figure 45 Top view of head showing linear features (left). Nostrils and lip-like features (right).

Correcting Geometrical Distortions

Because the MGS image was shot about 45 degrees off-angle it appears distorted — foreshortened in the direction of view. Late in the afternoon on April 6, JPL's Timothy Parker posted another image on the Internet[22]. This one was a geometrically stretched version of the original MGS image to correct for foreshortening — to simulate what the Face would look like from above.

---

[22] http://mars.jpl.nasa.gov/mgs/msss/camera/images/4_6_face_release/index.html

Image interpreters employ a method known as orthorectification to transform an image that has been acquired obliquely (off angle) to make it appear as if it were taken directly overhead. Only if the terrain is flat, however, can simple stretching be used compensate for the foreshortening in the direction of the observer.

Because the Face is not flat, JPL's correction actually distorted the Face even more (Figure 46). In their geometric stretch, the internal structure of the Face is pushed off to the wrong side in place by up to 400 meters, making it look much less symmetrical and face-like than it actually is.

To show this, I orthorectified the MGS image, using the same Viking-derived elevation model mentioned earlier to properly account for the shape of the Face. As a reference for comparison, I first orthorectified Viking frame 70A13, which was taken about 12° off-nadir. In comparing my orthorectification of the MGS image and JPL's simple stretch against the orthorectified 70A13 image of the Face, it was obvious that my rendition was a more accurate geometrical representation of the Face than JPL's (Figure 46)

Again, whether they just wanted a quick-and-dirty image for the public and the media, or whether there was another reason, JPL's 'geometrical-correction', like their earlier 'contrast-enhanced' image, was a gross misrepresentation of the data — distortions the scientific community and the media seemed to accept without question.

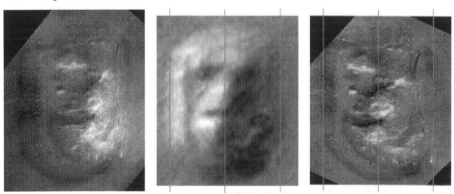

Figure 46 JPL's geometrically-stretched MGS image of the Face (left). Orthorectified 70A13 (middle) and MGS image (right). Lines indicate approximate horizontal locations of left and right bases, and centerline.

Announcing Our Findings

In the 1976 Viking images, the impression of a face was striking. Illuminated from below, the new MGS image seemed, at first glance, to be less remarkable. But on closer inspection, most of the features seen in the Viking images of the Face turned out to present in the new MGS image. The X-shaped pattern above the eyes, barely visible in the Viking images, is clearly resolved into a pair of linear features by MGS. In a computer generated perspective view of the top of the Face, these features appear to be cut into the formation, leading from the platform down to the ground below. Additional features, which look like nostrils and lip-like structures, were also found. Like the 'X' feature, they are located on either side of the centerline, the left-right axis of symmetry, of the Face.

Malin once said that to validate our hypothesis, "one must find something substantially different or unanticipated." It seemed that we had. Maybe the Face wasn't a natural landform after all.

The previous fall, several of us had submitted abstracts of our work on the older Viking images to the American Geophysical Union for presentation at their May meeting in Boston. But now, with the new MGS images, and after having spent the last couple of months studying this new image of the Face, we decided to present the results of our analysis to the scientific community. John Brandenburg and I prepared a poster presentation entitled: "Analysis of Unusual Martian Surface Features: Enigmatic Geology or Archaeological Ruins?"

Away on business, I was unable to be in Boston on the day of the presentation. John Brandenburg and Vince DiPietro were there along with several others from SPSR. Also there was Arden Albee, MGS's Chief Scientist, who McDaniel and I had met with a few months earlier. Albee was visibly disturbed by our conclusion — that the evidence in this new image sustained our original hypothesis the Face was artificial. A bizarre exchange followed, which, according to Brandenburg, went something like this [1]:

> Albee: "You say it's artificial."
>
> Brandenburg: " No. We only said it appears artificial based on our best analysis. It does appear to have eyes, a mouth and a helmet. Look, it appears to have two nostrils, too. Those were not visible in the Viking images."
>
> Albee: "No, you say it's artificial."
>
> Brandenburg: "No, the SPSR has always conceded that the face might be natural. Your people on the other hand, have never, not for an instant, admitted it might be artificial."

DiPietro then pointed out to Albee how the 'effective resolution' of the image was lower because of the very narrow range of gray levels due to the haze. Albee's temper flared, "You don't know what you're talking about!" DiPietro held his ground, "Yes, I do...." where upon Albee shouts at DiPietro, "Are you calling me a liar?" and raising his fists screams," I'll deck you, by God!"

Why had our findings evoked such a strong emotional response from, of all people, Albee? It seemed we had struck a nerve. Was it simply our tenacity, our refusal to abandon the Cydonia hypothesis despite all that JPL had done to 'scotch' the matter once and for all, or was there something more?

Around this time I received an email from the *Journal of Scientific Exploration* asking what I thought about the new image of the Face. I replied that despite all of the negative press we had found additional features suggesting the Face was artificial. I sent them a draft of a paper expanding on the AGU presentation. Previously, JSE had published Brandenburg, DiPietro, and Molenaar's paper on their Cydonia hypothesis, and my paper reviewing the evidence for artificiality in Cydonia. This time, however, its editor, Bernie Haisch, would not even consider a new paper on what we had found, stating in an email that, "it would undermine the scientific credibility of the Journal to publish your claim about the new MGS image still supporting a face on Mars. It is way too big a stretch."

What JSE did instead was invite David Pieri, a member of their editorial board, to write a paper on the new Cydonia image of the Face. Concerning the original photograph of the Face, Pieri states in his article [2]:

> Both Viking teams, and the public at large, were absolutely entitled to see faces, letters, or whatever else came to mind, as we looked at these pioneering images. There is an ancient and time-honored tradition of man anthropomorphizing nature, probably beginning with the Man in the Moon. It was good fun, and that was the spirit that moved us at the time. Little did we imagine the ruckus that would follow, and the 'cottage industry' that would develop as some observers astonishingly elevated the humble crumbling mesas and buttes of Cydonia to the status of putative intelligently sculpted artifacts. Those of us who had first looked at the images could see no logical basis for these hypotheses, even though many of us (including me) were then, and are now, strongly predisposed toward the SETI investigations. Finally, the ludicrous allegations of conspiracy and data suppression that followed over the years were particularly galling and scurrilous. Those of us on the Viking orbiter and Lander Science Teams would have liked nothing more than to trumpet to the world the discovery within our data of the evidence of an extraterrestrial civilization. But we just didn't see it. I firmly believe that someday—in this solar system or in some other—man will inevitably encounter such evidence, but, for me, the Face on Mars isn't it.

The rest of his paper addressed the geology of the Face and other landforms in the region. As in Sagan's "Man in the Moon" piece for *Parade*, and Malin's web page, Pieri makes no mention of any of our work, giving the reader the impression that there is no scientifically credible basis to our claim. In a response to Pieri's paper, published in the same issue of JSE, McDaniel and others wrote

> Pieri gives the impression that the facial appearance of the object is known to be an illusion of lighting. This interpretation was long ago refuted in a peer-reviewed article in which a three-dimensional model was derived and shown to produce the appearance of a face over a wide range of lighting and viewing conditions [my 1988 *Applied Optics* article]. We note that none of the Viking image team scientists or other critics of the Mars anomaly research have ever responded to this article in kind, that is, in a peer-reviewed journal, or elsewhere. Furthermore, the appearance of the object as shown in the photo taken in April 1998 by Mars Global Surveyor

(MGS) was accurately predicted in advance on the basis of this same 3-D model. Subsequent analyses of the recent image show numerous points of correlation between this and the earlier image, despite different lighting conditions... Even Carl Sagan in his 1995 book *The Demon-Haunted World* admits "There was an unfortunate dismissal of the feature by a project official as a trick of light and shadow" [3].

In one of his illustrations, Pieri 'drapes' the new MGS image over Malin's 3-D height map of the Face derived from the Viking imagery. The result was an extremely distorted rendering of how the Face might appear from the southwest. Again, on seeing this, an uninformed reader would have to agree with Pieri that there is nothing unusual about this landform.

According to their charter, the aim of the JSE is to "advance the study of anomalous phenomena." Clearly something else was at work here. Why else would a journal that had previously been open to the possibility of planetary SETI suddenly change their mind? Why would they refuse to allow new scientific results supporting earlier claims to be published, and at the same time, invite the opposition to publish a paper — one that does not properly reference the literature (not even to previous papers published in JSE), contains erroneous and misleading information, and make comments about conspiracy theories and 'cottage industries', issues irrelevant to the scientific issue at hand?

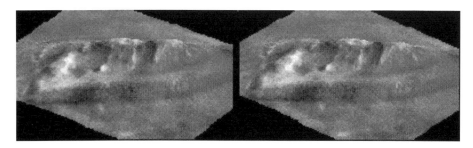

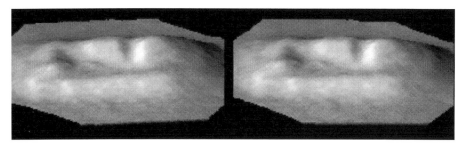

Figure 47 Simulated stereo view generated from the MGS image reprojected as it might appear from the southwest (top), and a similar view of the 70A13 data shown for comparison (bottom).

Other Anomalies in Cydonia

As we examined more of the MGS imagery it became obvious that any further evidence of artificiality would likely be extremely subtle. Although we were not finding Malin's roads and buildings, there were other features that cried out for explanation.

In comparing the April 6 MGS image (number 22003) with the earlier Viking 70A13 image over the same area, it seemed that the entire northeast quadrant of the D&M Pyramid should have been imaged at the bottom of the MGS strip. Because of haze and lack of shading, the pyramid was difficult to find. It was also hard to see because so little of it was visible due to the camera angle. At a 45-degree angle, a flat surface appears shorter by about 30%. Because MGS was looking roughly eastward toward the pyramid, the foreshortening of its northeastern side, which is sloped away from the camera, was even greater.

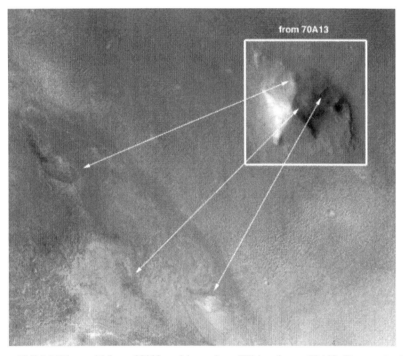

Figure 48 D&M Pyramid from 22003 and inset from Viking frame 70A13. Due to the MGS camera geometry, the shape of the D&M Pyramid is severely distorted.

In the older Viking images, the three illuminated faces of the D&M appear to be relatively flat with well-defined edges in between. Buttress-like structures at the base of several of the edges are also evident. In the MGS image (Figure 48), the edge between the northeast and northwest faces resembles a 'spine' running from the apex of the D&M down to the ground. At the base of the spine is a circular depression, possibly an opening. A dark blotch emanates from this circular feature and leads into a sinuous channel that runs off to the east. Due to the lack of shadows anywhere else in the image, it is unlikely that this feature is a shadow. One possible explanation is water leaching out of the D&M, darkening the soil as it flows to a low point in the terrain.

Two more pictures of Cydonia were obtained in April 1998. One, taken on April 14, caught the western edge of the City; the other, on April 23, captured the City 'Square', Starfish Pyramid, and a small part of the Fort. No significant anomalies are apparent in the second Cydonia image. The third image taken on April 23, provides an excellent view of the Starfish Pyramid, with the sun high enough in the sky to illuminate all of its sides (Figure 49). The high-resolution MGS image shows its five sides are divided by spine-like edges similar to those on the D&M pyramid[23]. A partial image shot from directly overhead the following year shows these spines are actually quite straight (Figure 50).

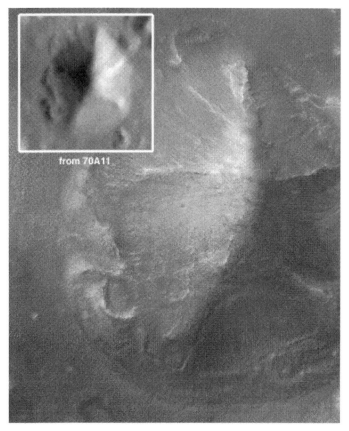

Figure 49 Starfish Pyramid from 25803 and inset from Viking frame 70A11. The view is from the north.

---

[23] Curiously, when viewed from the north, the five spines of the Starfish Pyramid resemble the shape of the five-pointed Egyptian star, which can be seen, for example, on the ceiling of the temple at Karnak.

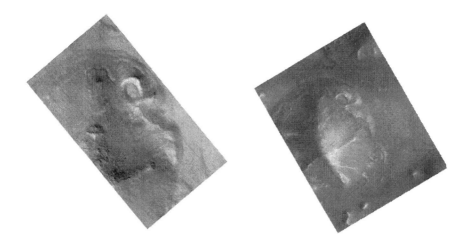

Figure 50 June 1999 image of portion of Starfish Pyramid acquired from almost directly overhead (left). Image has been rotated to better show straightness of visible edges. Compare this to the April 1998 image (right) acquired about 30° from vertical. Like the Face, the true shape of the Starfish Pyramid is distorted because of the camera geometry.

Several of the mounds analyzed by Crater and McDaniel are in the April 23 image. Up close they seem ordinary, yet the regularity of their layout remains unexplained. Near the top of the full image strip, several areas containing even smaller mounds or bumps — some only 5 to 10 meters in diameter — can be found. Although they are not aligned in any discernable pattern, the spacing between seems uniform[24]. A strange triangular feature that looks like an insignia was also found in this same general area (Figure 51).

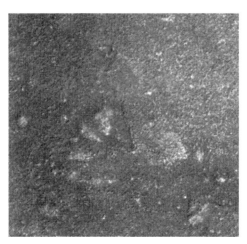

Figure 51 Triangular feature near the top of the April 23 MGS image (JPL/MSSS).

---

[24] Regularly spaced bumps similar to these and oriented in the same general direction as the crustal dichotomy were later found in a different part of Cydonia (image number SP249604) [4].

Aside from the Face itself, perhaps the most intriguing discovery was that the Face Starfish, and D&M Pyramid all have subtle linear and/or rectangular features at their bases (Figure 52). Recall that Erjavec found the Face to lie in a completely different terrain than the two pyramids. Both Erjavec and Pieri agree the Face shows evidence of water erosion. Yet the Face, which is morphologically different from the two pyramidal features, seems to have the same unusual pattern present in its side.

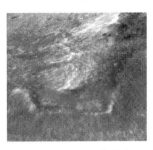  

Figure 52 Geometrical features at the base of the Starfish Pyramid (left), Face (middle), and D&M Pyramid (right).

NASA's Promise

After the initial set of targeted observations of the Face and City were made in April 1998, JPL seemed to loose interest in Cydonia. On November 11, 1999, SPSR President, Horace Crater, wrote to Carl Pilcher to find out why JPL did not seem to be honoring their pledge to image the Face and other features on an ongoing basis as had been promised. Noting that an imaging opportunity was missed a few months earlier, Crater wrote:

> When we met with you in 1997, you assured us that it would be NASA policy that during the mapping mission new images of the several suspect objects in the Cydonia region would be taken on every pass over the area, consistent with the safety of the craft. This promise was subsequently confirmed in a phone conversation with SPSR Vice-President Dr. David Webb by you as (in your own phrase) "official NASA policy." It was also confirmed in a telephone conversation with Prof. Stanley V. McDaniel with reference to the accuracy of his report of the meeting.
>
> Last July we wrote to MSSS (with a copy to you and Mr. Cunningham) informing them that our tracking sources had informed us that an imaging opportunity for the Face would occur in late August. Yet no image from that orbit has been released. If there were such an image, the official Data Release Policy would have dictated that by now it should have been released (that policy dictates that images of public importance should be released in seven days). So we must suppose that the opportunity, contrary to your stated policy, has been missed. This point is particularly important in the light of what was in fact released during that time period, which I will explain below.

What has developed during the period following the inadequate 4/98 image of the 'Face' appears to be a concerted effort to convince the public, on the basis of that flawed data, that the issue has been settled and furthermore to subject the hypothesis, and by implication those who support it, to ongoing ridicule. The resurgence of this policy of ridicule (and at one point blatant intimidation) threatens to destroy public confidence in NASA's earlier stated policy.

The first item in this seemingly propaganda-like pattern is the initial release of the MGS 'Face' image in April 1998. The processing of the image by MSSS was such as to render the feature flat, and combined with the grayscale limitation rendered it virtually featureless. Space-related magazines and news articles subsequently used this same image to cast derision on this topic.

The second item we note was the unfortunate occurrence at a poster session at the May 1998 AGU meeting, when a prominent Cal Tech planetary scientist physically threatened one of our SPSR members when the latter pointed out to him the lack of grayscales on the 4/98 Face image.

The third item is the action of Dr. Michael Malin, the Chief Scientist at MSSS, who was witnessed by one of our members to publicly apologize (while showing an image of the Face during a 1-hour invited presentation to the Division of Planetary Sciences of the American Astronomical Society on MGS results) to the planetary scientists for "demeaning science" and "wasting public funds" in taking these pictures. He insisted they were ordered by NASA over his protests.

Fourth, in a recent magazine article ("Getting the Picture," in, *Air & Space*, Vol.14, No. 3, August-September 1999, pp. 22-29, an aviation magazine put out by the Smithsonian Institution) by Henry S.F. Cooper, Jr. appears the statement: "If the picture of rocks on the plains of Elysium is Malin's favorite Global Surveyor image so far, his least favorites are the ones NASA ordered him to take of the so-called Face on Mars. In the new, high-resolution views, what had looked like a face in the old Viking pictures turned into what scientists had always suspected it was: just a jumble of rocks and outcrops. According to Malin, it cost $400,000 to take the new pictures. There were other targets that could have been viewed on the same orbit, including volcanoes on Elysium that would not likely come into view again. "Does the government spend money on ghost research?" Malin asks. "Or looking for the lost continent of Atlantis? I think the Face was a kind of stupid thing to spend money on."

This complaint about the cost to him (and through him the taxpayers) of $400K to take those 4/98 images of the Face and other Cydonia landforms is particularly disingenuous since just this past August MSSS took an image of the Crater Galle (the so called Happy Face Crater) with comments indicating it had little scientific value. (By the way that image was taken in the same month that one of the Cydonia Face was bypassed.) So we find that instead of taking a crucial image that would contribute greatly to the resolution of the issue, Dr. Malin spends public funds to once again resurrect the old 'Happy Face' argument, which we and others have shown time and time again is a *non sequitur* and serves no purpose other than one of propaganda and ridicule.

In July of this year a virtual replica of one of the 4/98 images of Cydonia (not the Face) was released by MSSS. Although displaying an interesting geological feature, this image did not image any of the three major features that bear on the question of possible artificiality. We very much regret that the August opportunity for imaging the Face was by passed, especially since it would have been a near vertical shot that would have an important bearing on that question. Another opportunity for imaging the Face will come in November 21 according to our sources. We hope it will not be missed again.

Two years after MGS reimaged the Face in April, 1998, MSSS released a handful of new images over Cydonia. Two were taken in August and November 1999, and covered the Fort in its entirety. Another, acquired in July 1999, captured most of the Tholus. Taken later in the mission after the spacecraft had transitioned into the mapping orbit, these new images are looking almost straight down, virtually eliminating the geometrical distortions that made it hard to interpret the first set of MGS images. In addition, the timing of the orbit had shifted so that the spacecraft was imaging Cydonia around 3 PM local time — closer to the time the Viking images were acquired. This made it easier to visually compare the two sets of data.

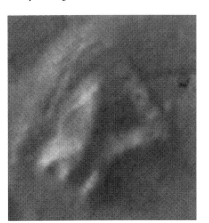 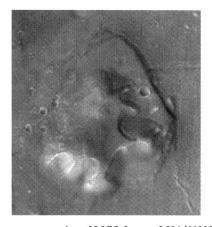

Figure 53 Fort from Viking 35A72 (left) and from a composite of MGS frames M04/01803 and M09/05394 (right).

In Viking frame 35A72, the Fort appears to consist of three straight sides enclosing a triangular inner space. The topography of the Fort, estimated from 35A72 using shape from shading, shows the feature to be roughly delta-shaped with the northeast side missing. There are peaks at the northern and southwestern vertices and an L-shaped depression just inside the northeastern edge. The impression of a three-sided object in the Viking image was created by the sun reflecting off the northwestern and southern sides and off the terrain sloping up to the northeast out of the L-shaped depression. With the sun to the northwest, the shadow cast by the northern peak into this depression created the effect of a triangular inner space. Illuminated from the south, the Fort loses its geometrical appearance in the MGS image (Figure 53). Up close it appears highly eroded like the Face and the other objects in Cydonia. Yet even though the straight edges are not as obvious in the new image because of the lighting geometry, they are still there.

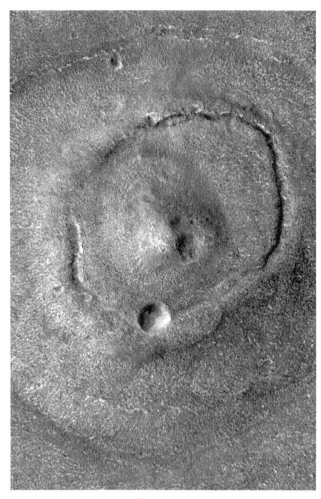

Figure 54 Enhanced MGS image of the Tholus (image number M03/00766)

The MGS image of the Tholus appears much like the earlier Viking images, but at a much higher resolution (Figure 54). The pit on the side of the structure looks like a small crater. The grooves might be fractures, or a channel formed by water runoff. The central peak is well defined, as is a shallow pit beside it..

Early in 2001, MGS imaged part of the west side of the Face from almost directly overhead. Although this image did not provide any new information about the Face, it did corroborate the accuracy of the orthorectified view computed from the April 1998 imagery. A portion of the Cliff was also imaged for the first time in 2001(Figure 55). The new image confirmed the sharpness and straightness of the central ridge running down the length of the feature.

How the Cliff was formed in such close proximity to an apparent impact crater remains a mystery.

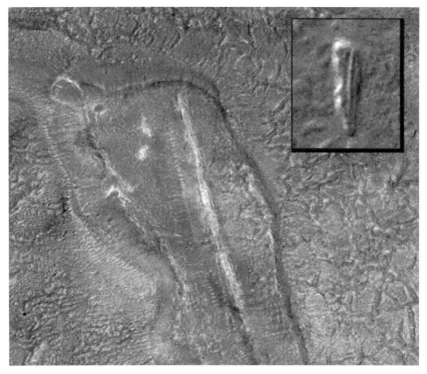

Figure 55 New image of the Cliff from (M18-00606) and original Viking image (inset).

Sources

1. John E. Brandenburg and Monica Rix Paxson, *Dead Mars, Dying Earth*, Element Press, Shaftesbury Dorset (England), 1999.
2. David C. Pieri, "Geomorphology of Selected Massifs On the Plains of Cydonia, Mars," *J. Sci. Exploration*, Vol. 13, No. 3, 1999.
3. Mark J. Carlotto, Horace W. Crater, James L. Erjavec, and Stanley V. McDaniel, "Response to Geomorphology of Selected Massifs On the Plains of Cydonia, Mars by David Pieri," *J. Sci. Exploration*, Vol. 13, No. 3, 1999.
4. Mark Carlotto, "Enigmatic Landforms in Cydonia: Geospatial Anisotropies, Bilateral Symmetries, and Their Correlations," *Sixth International Conference on Mars*, Pasadena, CA, July 20-25, 2003.

Nine — On the Threshold of a Dream

> There are more things in heaven and Earth, Horatio, than are dreamt of in your philosophy. — Hamlet, Act I, Scene 5.

As MGS continued its highly successful mapping mission, the controversy over the Face and other objects in Cydonia faded away. MSSS and JPL released thousands of images of the Mars. People from all over the world explored Mars via the Internet from the comfort of their homes and offices. They found all sorts of strange things: semi-transparent 'tubes', 'symbols' etched on the surface, dark streaks, unusual textured patterns, dark spots in dune fields, and many others.

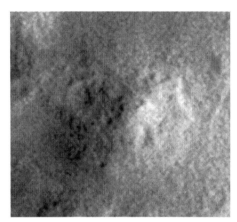 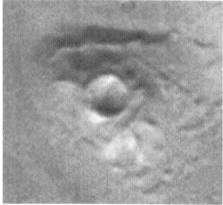

Figure 56 Rectangular feature on a teardrop-shaped island in image SP243304 (left). Note its similarity to the 'Bowl' from Viking frame 70A10 (right). (JPL/MSSS)

More geometrically-shaped objects were located. I found a small rectangular feature on a teardrop-shaped island (Figure 56) in an image taking during the Science Phasing Orbit (SP243304). This feature is similar to the 'Bowl' in Viking frame 70A10. In MGS frame M0701415, Mars researcher John Levasseur discovered a four-sided pyramidal formation with straight edges about 2 km across. Others too have found pyramidal objects.

Features that look like small letters and symbols etched into the surface were seen in several images over Cydonia. These features are present in the raw imagery data, which was posted on the Internet within a hours after receipt from the spacecraft. Neither imaging artifacts nor noise, they are a mystery.

Rippled patterns have been seen in a number of MGS images. Hoagland, Van Flandern, and others believe these patterns are semi-transparent tubes (Figure 57). Under certain viewing conditions they do seem to be tube-like structures, but after becoming better acquainted with the imagery, it is clear that they are dunes. Nestled inside narrow winding valleys, perhaps they are the fluvial remains of ancient streambeds — evidence for once-running water on the Martian surface.

Evidence of Water and Primitive Life on Mars?

One of the more puzzling discoveries made during this period were dark-colored elongated streaks or stains on the surface seen in numerous images. Scientific opinion is split on these features. Some contend the streaks are darker material under the surface that is being eroded away and transported by the wind. Others believe the streaks may indicate the presence of liquid water seeping out of the ground and flowing downhill, causing the light dry soil to darken. In a paper presented at the *2001 Mars Society Conference* entitled "A Study of Mars Global Surveyor (MGS) Mars Orbital Camera (MOC) Images Showing Probable Water Seepages," Efrain Palermo, Jill England, and Harry Moore argue for the latter interpretation. Their argument rests on a number of unique characteristics of these features [1]: 1) The streaks generally emanate from a point and

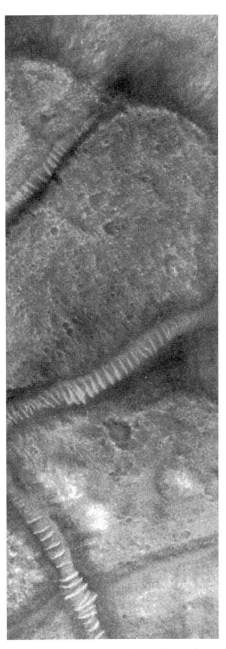

Figure 57 Features that have been interpreted by some to be tube-like structures from MGS image M1104220 (JPL/MSSS).

fan out in the down-slope direction. 2) Differences in coloration suggest changes of their composition with age (some seeps are lighter in color than others and even lighter than the underlying terrain), with darker seeps invariably overlaying lighter ones. The presence of light-colored streaks cannot be explained by wind action but "might be attributable to salts and minerals originally in solution with the seeping fluid and deposited on the surface as the liquid evaporated." 3) The similarity in the morphology of seeps to other seemingly water-related surface features such as gullies.

The seeps are geographically distributed around the Tharsis rise under which some believe vast reserves of water may exist. Analysis of a pair of images taken five months apart show significant change, suggesting the possibility that liquid water may exist at or near the surface of Mars today.

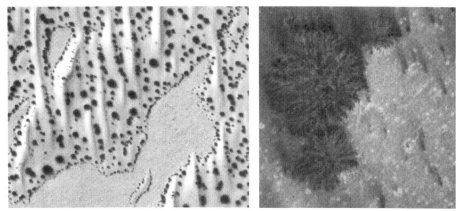

Figure 58 Dune spots thought to be colonies of dark-colored Martian microorganisms (left). A textured pattern from M0804688 (right). (JPL/MSSS).

Dark textured areas and 'dune spots' seen in a number of images over the south polar region, are two other puzzling discoveries. To science writer and visionary Arthur C. Clarke, the textured areas evoked the image of Banyan trees. NASA scientists lean more toward enigmatic geology — some natural process yet to be understood. In their paper, "Spider-Ravine Models and Plant-like Features on Mars — Possible Geophysical and Biogeophysical Modes of Origin," Peter Ness and Greg Orme examine these and other similar structures that vary seasonally in terms of shape, size and color. They consider the possibility that organic material, microbes, or even simple plants might exist within them [2].

András Horváth and Eörs Szathmáry have hypothesized that the dune spots might be colonies of dark-colored microorganisms [3]. They believe that during the winter, the soil below the spots is frozen, and that some form of ice or frost covers them. The organisms occupy a layer between the soil surface and the ice sheet. The organisms absorb sunlight and start to warm up at the end of the winter. As they thaw, the organisms find themselves in a liquid solute. In contact with the underlying surface, the microorganisms feed off nutrients in the soil. To test their hypothesis, Horvath and his colleagues have suggested

NASA land a rover in the south polar region so that it can observe changes in these features over the course of the year.

Many researchers who were first drawn to Mars by the mystery of the Face have made significant contributions in the search for water and primitive life on Mars, proof of which would be revolutionary discoveries in themselves. Yet the original questioned remained unanswered: Was the Face on Mars artificial or natural?

On The Symmetry of the Face

As part of the original independent Mars investigation, artist James Channon found the Face's platform and internal organization to be highly symmetrical. Analysis of the April 1998 image confirmed this. Using registered and orthorectified images derived from Viking frame 70A13 and MGS image 22003, I measured the positions of the left and right bases, left and right edges of the platform, and centerline of the Face. The centerline was established by running a line from between the 'X' pattern on the top of the head, down along the ridgeline defined by the points of highest elevation, between the two nostril-like features, and down through the lips. The measurements were then averaged to obtain estimates for the horizontal (lateral) position of the left and right bases, left and right edges of the platform, and the centerline. The distances between the centerline and the left and right platform edges were found to differ by about 8%. Using the distances between the centerline and left and right bases, the left side of the Face was a little more than 1% wider than the left.

The new MGS image confirmed the overall structure of the Face to have a high degree of lateral symmetry. But what about the internal details? Because of the extreme obliquity of the MGS image, very little of the right side of the Face was visible. But even though the right side had not yet been observed in its entirety, it did not seem to match the left. This apparent lack of symmetry has caused many to dismiss the Face as a natural geological landform. This kind of reasoning, like that of Sagan and Wallace, who, in their study of terrestrial satellite photos for signs of intelligent life on Earth, do not take erosion into account. In assessing possible artifacts (terrestrial or otherwise), it seems both reasonable and necessary to assume that a certain amount of degradation would occur over time. Erjavec and Brandenburg found what appear to be rills on several Cydonian landforms [4]. This is strong evidence these features were aerially exposed and erosion occurred through the actions of both precipitation and surface runoff.

Because artificial objects tend to be symmetrical, symmetry is a powerful feature for assessing the possible artificiality of objects in imagery. In their paper "Symmetry as a Continuous Feature," H. Zabrodsky, S. Peleg, and D. Avnir define the concept of 'symmetry distance' to be the minimum 'effort' required to transform a given shape into a symmetrical shape [5]. For a set of points, the effort is related to the amount each point must be moved from its location in the original shape to that in an assumed symmetrical shape. It seems reasonable that the amount of effort, in an erosional sense, required to transform a given landform into a symmetrical landform could be used as a means of assessing its artificiality.

Using this idea, I examined the orthorectified MGS image of the Face in order assess the possibility that, at one time, the Face might have possessed a much higher degree of internal symmetry, and that erosion could be responsible for its present condition. Upon careful examination, it appeared that deviations from symmetry in the Face could be explained by a small number of erosional events. Specifically, the right eye structure could be covered by debris that slid, slumped, or toppled down from what could have been the right extension of the brow ridge. The edge of a debris flow would account for the bright region below where we would expect to see an eye in the image. A debris flow would also account for the light region seen in the Viking image, just below the dark area that gives the impression of a right eye.

Another obvious deviation from facial symmetry is the light-colored crescent-shaped structure below and to the right of the chin. The shape suggests a dune, perhaps created from windblown material deposited on the leeward side of the Face. Light-colored material on the lee of an object is indicative of dust deposited by the wind. Above 40° N the prevailing wind direction on Mars is from the west [6]. As the left side of the Face appears darker, it is possible that lighter-colored material is being eroded from the left and deposited on the right side of the Face. Mass wasting is another possible explanation for this feature, whereby the material has either slid or slumped down from the right side of the central ridge.

The lack of strong shadows on the right side of the Face suggests the mouth does not extend across the face. However, on close examination subtle outlines are present. The right side of the mouth depression may be filled in with the same windblown material that forms the dune just below it.

That the Face might once have been much more symmetrical seemed a possibility (Figure 59). But in order to test this hypothesis another image showing the entire face in full illumination was needed. After more than three years the east side of the Face remained a mystery. No one had yet seen what was there, hidden in the shadows.

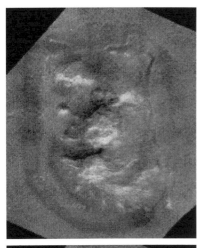

Figure 59 Notional reversal of the erosional process: Orthorectified MGS image (left). Speculative reconstructions follow: Reversing possible mass wasting event raises right eye (below left). Removing material possibly deposited in mouth reveals right extension (below right). Rubble on left side which may have slid down from nose ridge of face removed (bottom left). Dune formed from windblown material deposited on lee side of face below right side of chin removed (bottom right).

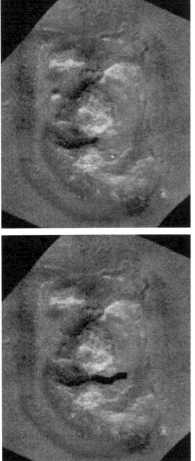

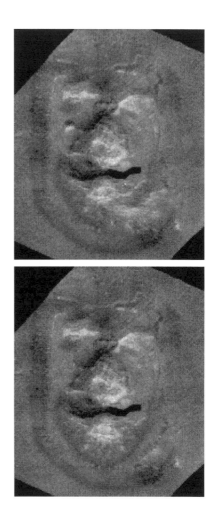

A Legal Ultimatum

Lack of new imagery of the Face and other Cydonia landforms prompted an organization known as FACETS (Formal Action Committee for Extra-Terrestrial Studies), headed by David Jinks, to issue a legal ultimatum to NASA to rephotograph these objects. In a letter dated 16 March 2001, written to NASA Administrator Dan Goldin on behalf of FACETS, attorney Peter Gersten states[25]:

> Though the MGS officially completed its primary mission earlier this year, coincidentally on the same day the latest Cydonia images were released, NASA has approved an extended MGS mission through April 2002. During this time the Mars Orbital Camera, freed from the constraints imposed by its previous global mapping mission, now can be routinely pointed at specific features on Mars.
>
> Thus, the opportunity to conclusively determine whether there was a former technological civilization on the planet Mars still exists. Higher resolution images of these previously identified, inexplicable structures — and at a different sun angle — is essential to this determination.

Continuing, Gersten requested that NASA release all imagery acquired over Cydonia to date, that the Face be targeted and any new images released immediately, and imagery over five additional areas be acquired.

NASA's reply was prompt. In a letter dated 11 May, Associate Administrator for Space Science Edward J. Weiler replies that "NASA has fully and openly distributed by means of public web-sites all images obtained of the Cydonia 'face' feature under question." He goes on to say that:

> None of the images acquired to date by the MGS/MOC system have been withheld and indeed, several recently (April 8, 2001) acquired images, including stereoscopic coverage of the Cydonia feature under question, have been released via multiple public web sites.

The fact that a new image of the Face had been acquired the previous month was news to everyone. For some reason, though, no one could find it on the web. Two weeks later, the image finally appeared. In a letter dated 5 June Gersten asks Weiler:

> Can you please explain why your letter dated May 11th stated, in several places, that the images were posted on the Internet on April 19th when apparently they weren't posted until May 24th ? Also, if this data was publicly accessible prior to May 24th, on which specific "multiple public web sites" were they?

Although a satisfactory answer was never received, we finally had that long awaited, high-resolution image of the Face in full sunlight, and it was remarkable.

---

[25] http://www.enterprisemission.com/letter.htm

Analysis of the April 2001 Image of the Face

Designated E03-00824, the April 2001 image, like previous ones, showed the Face to be a highly-eroded, asymmetrical formation. It was finally clear the right side of the Face did not match the left. JPL's interpretation was predictable. In a article posted on the Internet entitled, "Unmasking the Face on Mars," James Garvin, chief scientist for NASA's Mars Exploration Program states[26], "It reminds me most of Middle Butte in the Snake River Plain of Idaho. That's a lava dome that takes the form of an isolated mesa about the same height as the Face on Mars."

Using the Mars Orbiter Laser Altimeter aboard MGS to measure the height of objects, "We took hundreds of altitude measurements of the mesa-like features around Cydonia," says Garvin, "including the Face. The height of the Face, its volume and aspect ratio — all of its dimensions, in fact — are similar to the other mesas. It's not exotic in any way."

When the image was acquired on April 8, MGS had to be rolled 24.8° to see the Face 165 km to the west. The resulting image was somewhat distorted from how it would have appeared if it had been shot from directly overhead. Like the April 1998 MGS image, this new image had to be orthorectified — reprojected to look as if it was acquired looking straight down — in order to accurately analyze its 2-D plan geometry and symmetry. I did this by first computing a digital elevation model from the new MGS image using the single image shape-from-shading technique discussed in Chapter 5. SFS provides a high-resolution DEM that is precisely registered to the image. The DEM was then used to orthorectify the image.

One might wonder why I did not use the MOLA measurements to orthorectify the MGS image. Although MOLA can measure the heights of objects with a vertical precision of 20 to 30 cm, its horizontal resolution is very poor, about 150 meters per pixel — not detailed enough to accurately orthorectify a surface feature with complex internal detail such as the Face. But more importantly, the MOLA data cited in Garvin's article turned out to be the wrong data! Mars researcher Lan Fleming discovered that the data they used was not over the Face but over another mesa several kilometers away. After finding the correct MOLA data set for the Face, Fleming showed it was consistent with my SFS-derived DEM [7].

Using an algorithm developed by G. Marola to accurately locate axes of symmetry of objects in images [8], I determined that the lateral (left-right) and transversal (top-bottom) axes of symmetry of the Face intersected at a point at the center of a circular feature in the middle of the Face. In his article, Jim Garvin included a 'trail map' for the benefit of future Martian hikers exploring

---

[26] http://science.nasa.gov/headlines/y2001/ast24may_1.htm

the Face. The final destination on his map was the same circular feature that appeared to lie at the exact center of the Face on Mars.

As higher and higher resolution images of the Face were obtained, its face-like characteristics became less pronounced. In the April 2001 image there was no eye cavity on the left side of the Face, only 3-D features that cast shadows at certain illumination angles to create the impression of an eye. In the opinion of skeptics, the Face could not be a face because its details did not look face-like.

What I found next, however, seemed to render this criticism moot. In the April 1998 image, the circular feature mentioned above looked like a nostril. As shown in Figure 60, more accurate orthorectification now placed this feature (a) at the exact center of the Face — at the intersection of the horizontal and vertical axes of symmetry. The horizontal axis of symmetry passes through this and three other circular features (b-d) located along the centerline of the Face (Figure 61). This was not evident in the original image due to its distorted, off-nadir appearance. Continuing the analysis, I found that the locations of these features did not appear to be random.

The Face consists of a raised platform defined by a beveled edge. Examination of the shape of the bottom left edge of the base of the Face platform shows it to conform to a conic section, specifically to a segment from an ellipse, designated 'e1'. If we fit an ellipse to the base, the top of the ellipse passes through one of the circular features (c). A rectangle can be constructed whose left and bottom sides are tangent to the ellipse, and opposite vertex is in the center of the central depression (Figure 62). Three other rectangles of exactly the same size can be constructed using the center of the central depression as the common vertex (Figure 63). All four rectangles have an aspect ratio of 4 to 3; i.e., they are bisected by triangles whose sides are 3, 4, and 5 units long. The union of the four rectangles is also a 4-3 rectangle, designated 'EFGH'. Within this rectangle we can inscribe an ellipse 'e2'. The four points where this ellipse intersects the two diagonals of the rectangle define the four vertices of another, inner rectangle 'ABCD'. This is also a 4-3 rectangle, and is exactly one half the area of the outer rectangle 'EFGH'. These two rectangles neatly define the beveled edge on the left side of the Face, and bound it at the top and bottom (Figure 64).

As above, one can fit another ellipse 'e3' to the top left edge of the base of the Face platform. This ellipse passes through the center of another one of the circular features (b). The area of this ellipse is equal to that of the first. Within these two ellipses one can inscribe yet another ellipse 'e4' that passes through two of the circular features (b and c), and is tangent to the rectangle ABCD bounding the top surface of the Face platform. This ellipse is contained within a rectangle having an aspect ratio of 4/3. The likelihood that this particular configuration can occur at random is exceedingly small, about one in a million.

Although it can be argued that the likelihood of any particular spatial configuration of features occurring at random decreases as the number of features increases, the geometric complexity of the configurations will also tend to increase; i.e., they will not in general converge to a consistent, redundantly expressed geometry. Even though the Face is highly eroded, subtle evidence pointing to a simple, yet elegant, geometry such as this is difficult to explain as the result of a naturally-occurring random process on Mars, or anywhere else for that matter.

Further examination of the imagery provides additional evidence corroborating the highly symmetrical and geometrical structure of the Face. First, the top corners of the inner rectangle frame two rounded corners on the Face. The corners appear to have the same 3-D shape. Second, in our constructions, the right edge of the outer rectangle lies to the left of what appears in the image to be the base of the beveled edge. However, in comparing image intensity profiles taken from opposite sides of the Face, it is possible that the base does in fact coincide with the right edge of the outer rectangle. The impression that the base is to the right of where it should be, according to our model, is created by a build up of sand at the base. A computer-generated perspective view from below the chin (Figure 65), produced by mapping the orthorectified image onto the DEM, suggests the presence of sand dunes on the east side of the Face. As mentioned earlier, it is plausible that material has been eroded from the west and deposited on the east side of the landform by the prevailing winds. The depth of the sand may be sufficient to cover up some of the detail on the east side — including the right eye and right side of the mouth — thus explaining the difference in appearance of the two sides of the Face.

Using repeatable geometrical constructions based on clearly resolved features, the Face appears to conform to a simple architectural model based on rectangles having a long to short side ratio of 4 to 3. Earlier in the book we noted that the great 19th century German mathematician, Karl Gauss first proposed that mathematics, in particular the Pythagorean theorem — which relates the lengths of the sides of right triangles — be used as the basis for extraterrestrial communication. The right triangles that are formed by diagonally bisecting 4-3 rectangles are 3-4-5 right triangles. Knowledge of the 3-4-5 triangle predates Pythagoras and his followers by thousands of years. It was well known to the Chaldeans, Babylonians, and Egyptians.

How interesting that this, the simplest of geometries, seems to underlie the structure of this enigmatic object on the Martian surface.

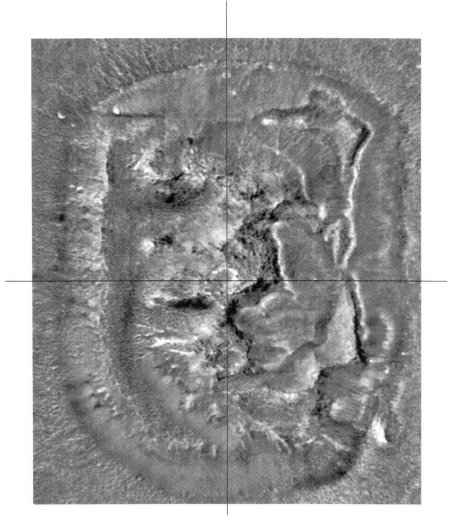

Figure 60 An orthorectified image with vertical and horizontal axes of symmetry shown. Note that the two axes intersect at the center of a small circular depression.

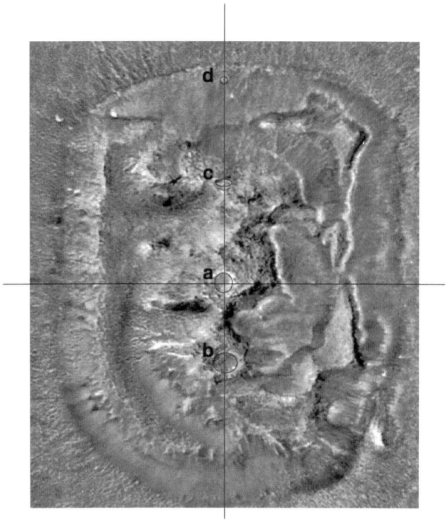

Figure 61 Four circular features (a-d) located along the lateral centerline (horizontal axis of symmetry).

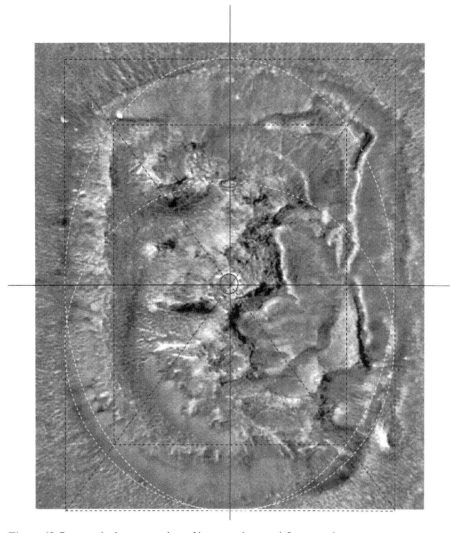

Figure 62 Geometrical construction of inner and outer 4-3 rectangles

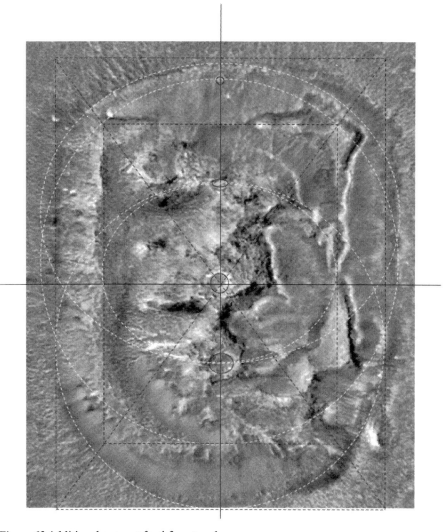

Figure 63 Additional support for 4-3 rectangles

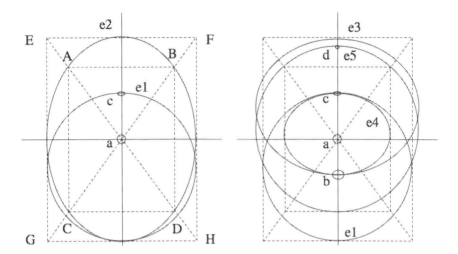

a) Circular features (a,c) along horizontal axis of symmetry. Outer rectangle 'EFGH' consisting of 4 smaller rectangles with vertices at E, F, G, and H, and sharing a common vertex (a). Ellipse 'e1' fit to bottom edge of platform used to construct outer rectangle. Ellipse 'e2' inscribed in outer rectangle. Inner rectangle 'ABCD' inscribed in ellipse 'e2'.

b) Circular features (a-d) along horizontal axis of symmetry. Ellipse 'e1' fit to bottom edge of platform used to construct outer rectangle passes through circular feature (c). Ellipse 'e3' fit to top edge of platform passes through circular feature (b). Areas of 'e1' and 'e3' are the same. Ellipse 'e4' passes through (b) and (c) and is tangent to inner rectangle. Circle 'e5' passes through circular feature (d), inner rectangle vertices 'A' and 'B' and is tangent to bottom of 'ABCD'.

Figure 64 Geometrical analysis of Face based on analysis of previous sequence of images

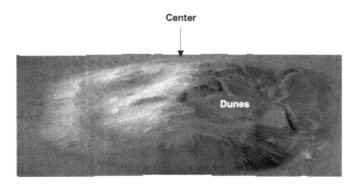

Figure 65 Perspective view of the Face from below the chin, showing dunes on the east side.

More on Symmetry

In their book *The Cydonia Codex*, George Haas and William Saunders argue the face is asymmetrical by design, depicting a different facial representation on each side. Although the split-face hypothesis, which was first suggested by Hoagland, leads to some interesting artistic and cultural speculations, it ignores the long-term effects of erosion and deposition, which are known to occur on Mars. Given the enormous age of the Face, it is likely that these and other natural processes are responsible, in large part, for its current appearance.

The split-face interpretation is intriguing (Figure 66): a feline face with flattened nose and brow on the right, a hominid face with raised eye and brow on the left.

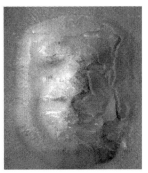 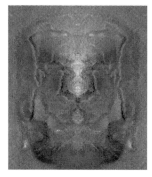 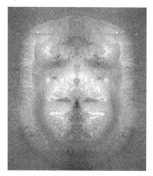

Figure 66 MGS image of the Face (left). Right side copied and flipped horizontally (middle) gives a feline impression. Left side copied and flipped horizontally (right) looks like a hominid, similar in appearance to our speculative reconstruction in Figure 59.

Although the split-face hypothesis does explain the current lack of internal symmetry in the Face, there are several problems with it: 1) Clear evidence of erosion (mass wasting?) on the lower right is ignored. Erosion would have obliterated the eastern extension of a mouth had one existed. 2) Clear evidence of deposition (sand?) on the right side below the feline eye down and to the right of the nose. This blanket of material appears to continue upward obscuring the right side of the brow, and possibly the right eye at the corresponding lateral position as well. 3) Asymmetry along the nose ridge that could be attributed to mass wasting on the left side of the Face. It has been suggested that the feature below the left eye might be debris that resulted from a landslide. A depression along the nose ridge above and to the east suggests a possible source for this material.

Figure 67 is a left-to-right transect through a digital elevation model of the Face just above the circular formation in the middle of the Face. It plots relative height (y-axis) as a function of lateral position (x-axis). The fall off in elevation is rather steep on the right side from the nose ridge down to the eastern side of the platform. The elevation profile on the left is more gradual, not unlike that of terrain that has collapsed and slid westward.

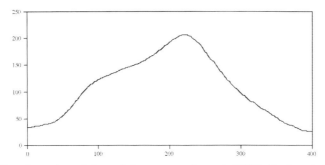

Figure 67 Elevations along left to right transect through DEM. Each unit in height (y-axis) equals about 1.15 meters (based on MOLA derived information). Each unit along the x-axis is 6 meters.

To test the hypothesis that material along the nose ridge may have slipped and slid down to the west, we assume that the terrain elevation surface was once symmetrical. For the transect above, a symmetrical profile (thick black line in Figure 68) is constructed from the original and flipped profiles (dotted lines). The sharp falloff in height on the right suggests that erosion and deposition were probably negligible on that side at this particular transect (vertical position). The hypothesized symmetrical height profile is based on the right side of the Face in its current state. On the left side, before mass wasting occurred, the elevation would have been higher toward the nose ridge (center) and lower as one approached the western side of the Face platform from the east (right to left).

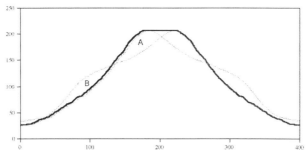

Figure 68 Hypothetical profile showing area where the distribution of material on the western (left) side of the face may have changed.

For this hypothesis to be physically plausible, the amount of material eroded from the nose ridge (A) would have to be equal to the amount of debris below (B). Figure 69 plots the difference in material between the solid and dotted lines on the left side in Figure 68.

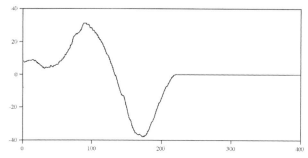

Figure 69 Loss and gain of material on the left side of the Face based on difference between curves A and B in previous figure.

Integrating (adding up) the differences from left to right shows the total amount of material is conserved (Figure 70), thus demonstrating this explanation is plausible. Whether or not it is the correct explanation is open to debate and further examination.

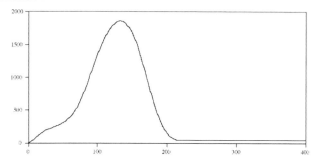

Figure 70 Adding up loss and gain shows amount of material is conserved on the left side of Face.

In contrast to the split-face hypothesis, the eroded-face hypothesis does not make any special assumptions about the Face. It only assumes that like its surrounding platform, which even today exhibits a high degree of symmetry, the Face was once a highly symmetrical object that has deteriorated over time.

Where MGS initially showed the Face to be much less symmetrical than we had originally thought based on the Viking imagery, more recent imagery from the Thermal Emission Imaging System aboard the Mars Odyssey spacecraft revealed a surprising new symmetry in the D&M Pyramid — one that was not at all obvious in the Viking data (Figure 71).

Measuring angles in Viking image 70A13, Torun identified a number of relationships in the D&M's internal geometry that suggested an underlying design based on tetrahedral geometry. At twice the resolution of the Viking data, more accurate measurements of these same angles in the THEMIS image seem to suggest a different architecture. If we split the pyramid along its axis of symmetry (Figure 72), in plan view, each half consists of a 30 degree isosceles triangle, a right triangle, and a equilateral triangle. In all, the D&M appears to be composed of five triangular facets — three are equilateral and two are right triangles. Equally intriguing is that the D&M's axis of symmetry lies roughly in the direction of the grid discovered by Crater and McDaniel in their analysis of the mounds, which is also aligned with the major axes of several other objects including the Face.

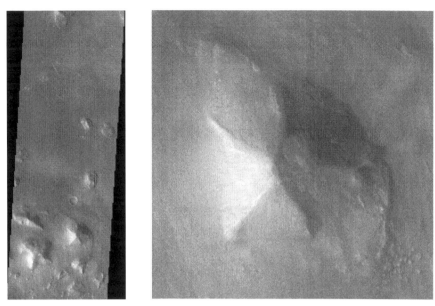

Figure 71 THEMIS image strip 20020413a (left). Image is 53.4 by 22.5 km in area at a resolution of 19 meters/pixel. Close up of D&M Pyramid (right). (JPL/ASU)

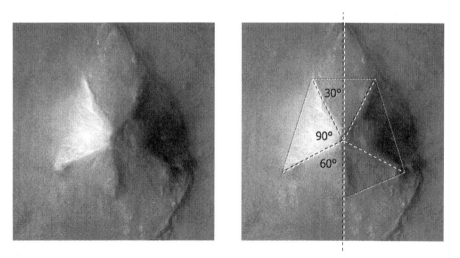

Figure 72 D&M Pyramid rotated so that its axis of symmetry is vertical (left). New geometrical model of the D&M Pyramid (right).

Sources

1. Palermo, E., England J., and Moore H., "A Study of Mars Global Surveyor (MGS) Mars Orbital Camera (MOC) Images Showing Probable Water Seepages," 2001 Mars Society Conference, Palo Alto, CA.
2. Peter K. Ness and Greg M. Orme, "Spider Ravine Models and Plant-Like Features on Mars - Possible Geophysical and Biogeophysical Modes of Origins," *J. Sci. Exploration*, Vol. 55 No 3/4, March-April 2002, pp 85-108.
3. Tibor Gánti, András Horváth, Szaniszló Bérczi, Albert Gesztesi and Eörs Szathmáry, "Dark Dune Spots: Possible Biomarkers on Mars?," *Origins of Life and Evolution of Biospheres*, Volume 33, Numbers 4-5, October, 2003.
4. Erjavec, J. and Brandenburg, J., "Evidence for a Paleo-Ocean Shoreline, Sedimentary Features and Water Erosion in Cydonia Mensae," AGU Spring Meeting, June 1-4, Boston MA, Abstract, P42A-10 (1999).
5. Zabrodsky, H., Peleg, S., and Avnir, D., "Symmetry as a Continuous Feature," *IEEE Transactions on Pattern Analysis and Machine Intelligence*, Vol. 17, No. 12, Dec. 1995.
6. Carr, M., *The Surface of Mars*, Yale University Press, New Haven CT, 1981.
7. Fleming, Lan, "Identification and Evaluation of the Mars Global Surveyor MOLA Profile of the Mars Face," *New Frontiers in Science*, Vol. 1, No. 1, Fall 2001 (http://www.lulu.com/content/1748865).
8. Marola, G., "On the detection of the axes of symmetry of symmetric and almost symmetric planar objects," *IEEE Trans. Pattern Analysis and Machine Intelligence*, Vol. 11, No. 1, Jan. 1989.

Ten — Toward a Synthesis of Science and Myth

> The point that stands out most clearly is the need for a greater humility in dealing with the philosophical, religious, and scientific issues so central to the debate [over extraterrestrial life]. Many of our finest minds have faced these issues; directly or indirectly, the message their writings seem most dramatically to demonstrate is that the ways of the universe and God are more difficult to discern than most inhabitants of our planet have been willing to recognize. — Michael J. Crowe

The debate over extraterrestrial life, in the words of Michael Crowe, "has produced hundreds of claims, thousands of publications, and millions of believers, but not as yet a single solid proof" [1]. Are the Cydonia discoveries to go down in history along with Lowell's canals, Gruithuisen's walled city on the Moon, and other claims that have fallen by the wayside, or do they represent the start of a new chapter in the search for extraterrestrial life, one that will finally lead to a definitive answer?

That the debate over extraterrestrial life has gone on for so long provides us with a rich historical context for evaluating the Cydonia discoveries. In the words of Spinoza, "If you want the present to be different from the past, study the past." In this chapter we begin by examining the nature of the evidence presented thus far as well as the arguments used on both sides of the controversy to see how well the Cydonia discoveries hold up to scrutiny.

The Cydonia Controversy and the Extraterrestrial Life Debate

Crowe's in depth analysis of the extraterrestrial life debate from 1750-1900 reveals a number of what he calls "recurrent fallacies and linguistic abuses" that have led to unfounded claims and heated arguments over extraterrestrial life. Some have to do with the nature of the arguments used, others with the numbers and probabilities used to back up an argument, and still others with the role of one's prior belief in the hypothesis under consideration. Let us consider these in the context of the Cydonia controversy.

The first has to do with the tendency to rely on analogical arguments. To reason by analogy is to assume that if two things are alike in some respects, than they must be alike in others as well. In science, analogical reasoning is accepted as a method of discovery, but is not, except under special circumstances, an acceptable method of proof. Initially, Hoagland used analogical reasoning to expand the search beyond the Face to other anomalies in Cydonia. On Earth, large structures tend to be near populated areas. Reasoning that if the Face is artificial, someone had to build it, he searched for a place where the builders could live, and in the process, found the City. Analogical reasoning led Hoagland to other anomalies in the area, and helped to define general criteria for characterizing artificial objects based on the characteristics of artificial structures on Earth; e.g., by their non-fractal structure, symmetry, rectilinearity, and others. Analogical arguments have not, however, been used as evidence. The claim that this collection of objects on Mars is artificial is not based on the observation that they resemble a city. Rather, as we discussed in Chapter 5, it is based on a number of intrinsic properties of the objects themselves such as non-fractal structure, symmetry, geometry, non-random location, and similarity in size, shape, and orientation to other objects.

Another stumbling block in the extraterrestrial life debate is in mistaking necessary for sufficient conditions. If 'A' is a necessary condition of 'B', then 'A' implies 'B.' For example, if 'A' represents 'life,' and 'B' represents 'water,' then 'life' implies 'water.' Because life cannot exist without water (as far as we know), water is a necessary condition for life. The presence of life implies the presence of water. However, water is not a sufficient condition for life since living creatures require other things to survive ('B' does not imply 'A'). Finding water does not prove the existence of life.

Carl Sagan's famous quote "Extraordinary claims require extraordinary evidence" has led some to demand that proponents of the Cydonia hypothesis produce proof of artificiality. In Malin's 1992 letter to Dan Drasin, he stated "Investigations of the 'Face' originally cited its 'bilateral symmetry' as proof of its artificial nature." Saying that a measurement like bilateral symmetry proves that an object is artificial is the same thing as saying that bilateral symmetry implies artificiality — that it is a sufficient condition. Malin tries to disprove the hypothesis that the Face is artificial by saying that because "the landform is clearly not bilaterally symmetric" it is not artificial. As we shall see later in the chapter, there are many ancient structures on Earth that are not symmetrical because of erosion. Lacking one piece of evidence, are we to conclude that they are not artificial?

The problem with Malin's criticism is that it was never claimed that symmetry or any other single characteristic is sufficient to prove that the Face is an artificial object. That bilateral symmetry, anthropometric facial proportions, non-fractal behavior and other characteristics would likely be observed if the Face were the result of intelligent design is to say that they are necessary conditions. This distinction between necessary and sufficient conditions is clear in the argument for artificiality, which is based on the idea of 'convergent evidence' — weak evidence from multiple independent sources tending to support one hypothesis over another. Although it is claimed that the sum of all the evidence favors intelligent design over geology, and that the weight of the evidence is strong enough to overcome a reasonable doubt, no one has claimed the current evidence is strong enough to prove that the structures are artificial. The question of artificiality will not be resolved until we land on Mars and examine these structures in person.

A related problem in logical reasoning is the use of what Crowe calls 'hypothetico-deductive' arguments — arguments based on showing that because a particular set of observations support a given hypothesis, the hypothesis must be true. Strictly speaking, only when all other hypotheses have been disproved can the hypothesis under consideration be considered proved. For example, changes observed in the tones and colors of Mars led 19th century astronomers to believe that Mars had water. The lines first seen by Schiaparelli led Percival Lowell and others to conclude that the Martians had built canals to save their dying planet; but of course, there was another explanation for the color changes. In Cydonia a variety of geological mechanisms have been proposed to explain these objects, from wind erosion to volcanism. Although it is plausible to explain individual features in terms of 'enigmatic' geology — by a coincidence of random events that conspire to produce an artificial looking object — it is a stretch to extend this line of reasoning to the whole collection of objects.

What makes this different from previous extraterrestrial life claims is that it is based not on one or two pieces of evidence but on many. The argument, which is a probabilistic one, is that as the amount of available evidence increases, the most likely hypothesis will dominate. In effect, it is the sum total of all of the evidence that effectively eliminates other possibilities. Never before has such a broadly based argument been made in support of an extraterrestrial hypothesis.

Crowe notes that in making a probabilistic argument for the existence of extraterrestrial life one must be careful to distinguish between inductive and theoretical probabilities. As an example of the use of theoretical probabilities, consider Frank Drake's famous equation[27] for estimating the number of civilizations in the Milky Way Galaxy whose radio emissions are detectable by us. It expresses this number as a product of seven terms:

$$N = R^* f_p n_e f_l f_i f_c L$$

where:

> R^* is the rate of formation of stars with a large enough 'habitable zone' and long enough lifetime to be suitable for the development of intelligent life,
>
> f_p is fraction of those stars with planets,
>
> n_e is the number of 'Earths' per planetary system (planets able to maintain the basic conditions for life as we know it),
>
> f_l is the fraction of those planets where life develops,
>
> f_i is the fraction life sites where intelligence develops,
>
> f_c is the fraction of planets where technology develops, and

---

[27] http://en.wikipedia.org/wiki/Drake_equation

L is the "lifetime" of communicating civilizations (the length of time such civilizations release detectable signals into space).

Moving down this list of factors, the values become increasingly more speculative in nature.

In contrast to the Drake equation and its theoretical probabilities, the case for artificiality in Cydonia is based on the product of empirically-derived probabilities, probabilities that one can actually calculate. As discussed in Chapter 5, using ground truth over terrestrial study areas it is possible to estimate one of the probabilities in the equation — the ratio of the probability that the fractal technique detects an artificial object given it is there divided by the probability that it detects an artificial object given it is not there. Although the ratio of probabilities (the 'weight' of each piece of evidence for artificiality) could, in principle, be estimated given a sufficient number of terrestrial archaeological case studies, for the sake of expediency, we assume that the other sources of evidence are similar in weight and independent of one another. Given an even chance of finding artifacts on Mars (i.e., given no prior knowledge or subjective bias either way), the probability these objects are artificial exceeds the probability they are natural by a factor of over 17 billion to one!

One has to wonder why, given its lack of success and highly speculative theoretical foundation, radio SETI continues to receive mainstream support from the scientific community while alternative SETI concepts such as the search for planetary artifacts and the study of UFOs are marginalized by the same institutions.

Returning to possible pitfalls, another problem involves the misuse of large numbers to justify an extraterrestrial assertion. According to Crowe, those who believe in extraterrestrial life have relied on the assumption that "among the countless planets orbiting the billions of stars in our galaxy, some at least must be inhabited." For example, Shklovsky and Sagan's Assumption of Mediocrity states that our galaxy contains a tremendous number of planets on which life could develop and evolve, and that there is nothing special about Earth to favor it over other planets. That man is the result of evolution on Earth leads to the assertion that there must be other intelligent life in the galaxy. Ironically, arguments of a similar nature have been turned around and used to dismiss the Face and other features in Cydonia as simply odd-looking rock formations. Geologist Harold Masursky once said that among the thousands of mesas in Cydonia he would have been surprised if none looked like a face. An anonymous NASA reviewer of Hoagland's book the Monuments of Mars uses a similar argument [2]:

> The author appears to have no appreciation of the enormous complexity of Martian geomorphology, which is actually more complex than that of Earth because of the random occurrence (in time and space) of major impacts, superimposed on an already evolved terrain. The Martian landscape as we see it reflects the influence of

at least the following processes: volcanism, tectonism, catastrophic flooding, mass wasting (with and without ground ice), freezing and thawing, wind erosion and deposition, fluvial erosion, perhaps glaciation, and impact. Furthermore, the Martian surficial geologic record is clearly much longer than that of Earth; instead of seeing landforms dating back not more than probably a few hundred million years, we see (on Mars) a montage of features probably going back billions of years.

In the same way arguments involving large numbers do not prove extraterrestrial life, these statements do not prove the Face has to be natural. They simply represent opinions, informed opinions, but opinions nonetheless.

Perhaps the most common mistake made in the SETI debate has been to assume that the question of extraterrestrial life is purely scientific. The inability to separate opinions (one's prior beliefs) from evidence has led to a great deal of misunderstanding over Cydonia. What one believes is almost never purely scientific. Steven Dick discusses the role of prior beliefs (metaphysics) in the extraterrestrial life debate. Concerning Kepler's strong belief that the Moon was inhabited he states [3]:

> Kepler's axiomatic demonstration of the existence of lunar inhabitants is, with the benefit of hindsight, a demonstration of the role of metaphysics in action. The order perceived by Kepler originated not in the minds of intelligent extraterrestrial beings, but in the mind of the observer himself, who imposed order where in fact there was none, or, at least, could not fathom the natural cause of the apparent order.

He goes on to say that "This type of argument was not unique to the seventeenth century or to the history of science; in a similar context at the turn of the twentieth century, the identical effect was at work when the astronomer Percival Lowell 'observed' canals on Mars."

The ability to treat evidence and prior belief explicitly and independently is essential to avoiding misunderstandings. Evidence has nothing to do with personal belief. In fact, under certain conditions as mentioned above the degree to which a piece of evidence supports a particular hypothesis (technically, the probability that a particular hypothesis is true given the evidence) can actually be computed. All one's belief does is to scale the probabilities up or down. What is important to realize is that as long as the prior probability of one's belief in the hypothesis under consideration is not zero (is not impossible), it is possible, given enough evidence, to determine in a quantitative manner whether it is true or false. For example, if we assume the likelihood of an extraordinary claim is one in a million, the sum total of the evidence for artificiality in Cydonia still far exceeds reasonable doubt by a factor of over one in 17,000. That so many pieces of evidence point to intelligent design challenges conventional explanations. Perhaps Greg Molenaar was right in invoking the principle of Ockham's Razor — that one should make the minimum number of assumptions in order to explain a phenomena — when he said: "If a mountain clump on Mars looks like a carved humanoid face, the most simple explanation may be that it is!"

Last but not least is the importance of empirical evidence. Lowell's claim that the canals were built by Martians trying to save their dying planet was based on two assertions: 1) Mars had water and, 2) because Mars had water, it had life. The existence of the canals and their Martian builders was later denied by empirical evidence — Maunder's demonstration of the tendency of the human visual system to see structure in cases where there was none, and spectroscopic data showing the atmosphere to be too thin to support water and the kind of life Lowell and others had imagined.

Curiously, in Cydonia the situation is reversed. The case for artificiality rests on the large body of empirical evidence summarized in this book. The planetary science community's negative position regarding Cydonia is based largely on two contentions: 1) the Face and other objects do not appear to be artificial, and 2) a suitable environment did not last long enough for advanced life to develop on Mars. Neither one has been established.

The amount of time required for intelligent life to develop in a suitable environment is unknown. Current scientific thinking that it takes 3.5 billion years may be mistaken, particularly in light of evidence in the fossil record that indicates life undergoes periods of rapid development followed by long periods of little or no change. That advanced life could have either developed during an accelerated time scale, or could have been introduced to Mars from somewhere else is not beyond the realm of possibility.

Critics claim the April 1998 and 2001 MGS images of the Face prove it is natural. While the images do show the Face is much more eroded than originally thought, there are definite indications of symmetry and geometry as we have seen. That these objects do not look artificial to planetary scientists stems from their expectation that they will either find obvious evidence of relatively recent habitation or nothing at all on Mars (e.g., Sagan and Wallace's extension of criteria used to detect signs of life on Earth to Mars, and Malin's scenario of finding either roads and buildings, or rocks, cliffs, channels, and dirt).

The possibility of finding ancient and thus highly eroded archaeology on another planet has not been previously considered, nor has the problem of distinguishing eroded architecture from geology by remote sensing. These objects in Cydonia along with recent discoveries on Earth — discoveries which defy explanation in terms of their size, level of sophistication, and age — are forcing us to face this possibility for the first time.

Extraterrestrial Archaeology

Terrestrial archaeology is defined as the systematic recovery by scientific methods of material evidence remaining from man's life and culture in past ages, and the detailed study of this evidence. If exobiology is branch of biology concerned with the search and study of extraterrestrial life forms, it seems reasonable to define extraterrestrial archaeology (or exoarchaeology) as the study of the material evidence (e.g., architecture, artifacts, etc.) of these life forms.

In contrast to the diversity of potential sources available in conventional archaeological studies, extraterrestrial archaeology must for now rely on remotely-sensed satellite data and the analysis of patterns and relationships in the data. According to archaeologist Kevin Greene, mathematical techniques are useful in analyzing data ranging from regional studies to the positions of individual artifacts. He states that it is often important to "determine whether scatters of sites or artifacts contain significant clusters or a regularly spaced pattern, or whether their disposition is purely random" [4]. He goes on to say that "When the distributions of more than one kind of artifact or settlement have been recorded, comparisons may be made to establish connections (correlations) between them..."

Crater and McDaniel have shown in their study of the mounds in Cydonia that a strong non-random statistical anomaly exists in their arrangement, specifically in the number of '19.5° triangles' that can be drawn between them. Recall from Chapter 5 that if you turn a tetrahedron so that its base is up (to the north) and place it inside a sphere, the 'latitude' of the three vertices containing the base is at about 19.5° N. Other references to tetrahedral geometry also exist; e.g., Torun's claim that the ratio of the surface area of a circumscribed sphere to that of the tetrahedron inside, roughly 2.720669, which he called e' is related to the position of the D&M Pyramid on the surface of Mars, specifically that the latitude given by the arctangent of e'/π passes through the D&M. Some believe that if these objects are artificial perhaps there is a message encoded in their locations and alignments.

But there is another interpretation of Crater and McDaniel's findings that leads to some interesting correlations with other observations in Cydonia. The '19.5° triangles' defined by the mounds include right triangles having the angles 90°, 54.75°, and 35.25°, and isosceles triangles with angles 70.5°, 54.75°, and 54.75°. McDaniel noticed that the vertices of these triangles coincide with the vertices of a rectilinear grid that is oriented at an angle 35.25° north of east.

A powerful archaeo-astonomical technique used to date ancient monuments on Earth involves the analysis of alignments with respect to the cardinal (compass directions) and directions of astronomical importance such as that of the summer and winter solstice. For example, to an observer standing at the center of Stonehenge on the first day of summer, the sun rises slightly to the right of the northeast axis of the monument. Because of changes in the obliquity or axial tilt of the Earth, the location of the summer solstice sunrise changes ever so slightly over time. The last time the sun rose in line with Stonehenge was around 2700 B.C., which is the date Stonehenge is thought to have been built.

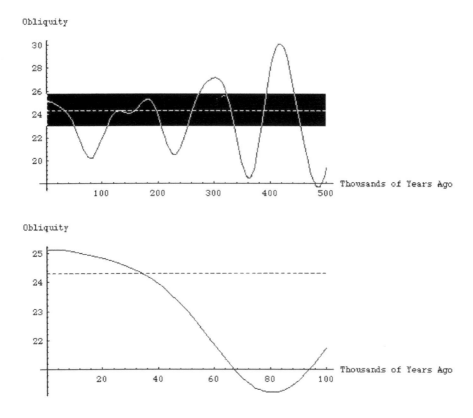

Figure 73 Predicted obliquity curves for Mars based on a model developed by Ward [5]. The dotted line in the top curve is the value of the obliquity (24.4°) required for the sun to rise 33.3° north of east on the first day of summer on Mars. The black band in the figure represents the range of obliquity values (24.4 ± 1.5°) that correspond to the uncertainty in the estimate of the orientation of the complex (33.3°± 2.07°). The bottom curve expands the first 100,000 years of the predicted obliquity curve. The earliest the solstice alignment could have been satisfied for the Cydonia complex was about 33,000 years ago.

As discussed in Chapter 4, early in the investigation Hoagland speculated that the City Square was a good vantage point to view the Face in profile. In measuring the angle of a line between the City Square and Face, about 23.5° north of east, the thought occurred to Hoagland that its direction might be solsticial and thus provide a clue as to how old these objects were. Based on his measurements and Ward's model [5] describing the changes in the obliquity of Mars over time, Hoagland estimated the sun rose over the mouth of the Face as seen from the City Square as early as 500,000 years ago. Although Hoagland's use of archaeo-astonomical techniques to date these formations on Mars has been criticized on the grounds that the objects have not been shown to be artificial, and that the sightline between the City Square and Face is somewhat arbitrary, he was the first to notice the City and Face appeared to be aligned with one another. In Chapter 5 we showed that the orientation of several of the larger objects in Cydonia, namely the Face, Fort, Starfish Pyramid, and a rounded formation west of the City all lie in the same general direction, about 31.8° north of east. That the orientation of the mounds is about the same as the larger objects in Cydonia to within a few degrees, well within measurement error, may be more than just an interesting coincidence.

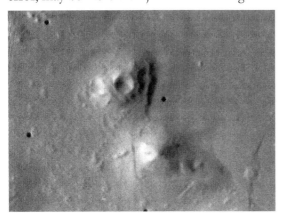
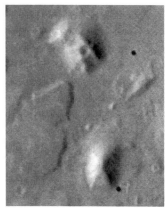

Figure 74 Other sites on Mars having the same configuration at the City and Face. Viking frame 70A10 west of the City and Face containing a bowl-like formation and nearby pyramidal object (left). Frame 70A01 further to the west contains similar objects (right).

Within this collection of objects, which all seem to point in the same north-of-east direction, lies the D&M Pyramid with one of its sides facing due south. Another interesting coincidence is that two other sites in Cydonia appear to have the same organization. The first, seen in Viking frame 70A10, is southwest of the City and Face. This site contains a circular formation situated on a raised rectangular platform (the 'Bowl') oriented in the same general north-of-east direction as the City and Face. Nearby is a pyramidal object ('B pyramid') with one of its faces pointing to the south. The second site, found by Ananda Sirisena, is farther to the west and contains a similarly aligned pyramidal object, which he named the 'King Pyramid' with a south pointing face, and another bowl-like formation called 'Fort Aetherius' oriented in a similar north-of-east direction. All three sites contain a pyramidal object that appears to be aligned to the meridian plus at least one other object oriented in the same general north-of-east direction.

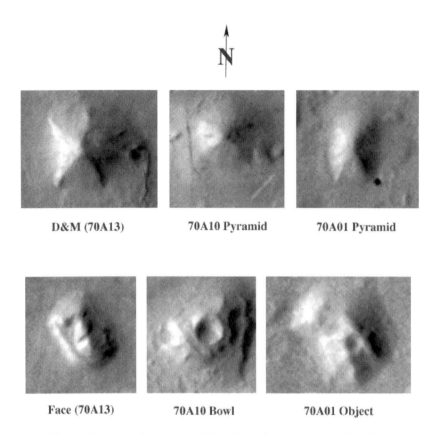

Figure 75 Three sites contain a pyramidal object that appears to be aligned to the meridian, plus at least one other object oriented approximately 33 degrees north of east. Shown above (left to right) are the D&M Pyramid from Viking frame 70A13, and pyramidal objects from 70A10 and 70A01. Shown below (left to right) are the Face, Bowl from 70A13, and object from 70A01.

Let us examine four possible explanations for these correlations. First, consider geology. As discussed in Chapter 4, Mars' crustal dichotomy is marked by a series of cliffs, faults, and escarpments that lie along a great circle inclined about 35° to the equator. That these sites lie in the same general direction of the crustal dichotomy in this part of Mars would seem to suggest the origin of these features is geological not architectural. There are however two problems with this explanation. First, the origin of the crustal dichotomy itself is not known. To explain these anomalies in terms of a geological feature that is not well understood is not a satisfactory explanation. Second is the fact that this oriented pattern exists over a wide range of scale, from mounds that are several tens of meters in diameter, to the Face and other formations that are roughly a kilometer in size, to the sites themselves distributed within a region on the order of hundreds of thousands of square kilometers in area [6]. No natural phenomenon like it exists, to our knowledge, either on Earth or on Mars.

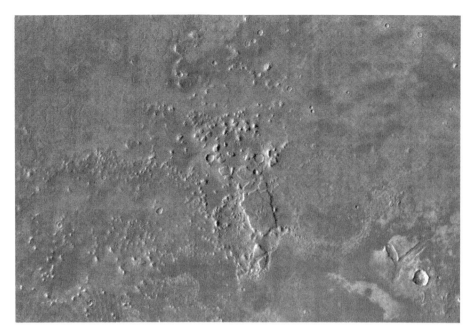

Figure 76 Composite of two Mars Digital Image Model images covering the area from 37.5-42.5°N latitude and 4.5-15.5°W longitude in Cydonia. Landforms in this part of Mars, including the City and Face (center) and other anomalies in two other two sites (left of center), appear to be aligned in the same general north-of-east direction. (JPL/USGS)

Three other possible explanations involve intelligent design at different levels of sophistication. During the independent Mars investigation it was observed that the features in Cydonia are near the 0 km datum, close to sea level. Although the idea that a large northern ocean once existed on Mars is in itself a rather controversial issue, it has gained considerable momentum recently with the publication of a paper by James Head and his colleagues entitled "Possible Ancient Oceans on Mars: Evidence from Mars Obiter Laser Altimeter" published in the December 10, 1999 issue of the journal *Science*. In it they state [7]:

> High-resolution altimetric data [collected by the MOLA instrument onboard the MGS spacecraft] define the detailed topography of the northern lowlands of Mars, and a range of data is consistent with the hypothesis that a lowland-encircling geological contact represents the ancient shoreline of a large standing body of water present in middle Mars history.

If these sites are the archaeological remains of an extensive complex of some kind, and that a large body of water once existed north-northwest of these features, it is not unreasonable that a pattern of development might have followed the general direction of the shoreline. In the American Midwest where the terrain is flat, roads and towns tend to be aligned north, south, east, and west. However, along the Eastern seaboard, the pattern of roads and towns,

163

particularly older ones, follow the shoreline. Could a similar pattern of development have occurred on Mars?

Another possible explanation for the alignment is based on the idea of polar wandering, first suggested by astronomer Thomas Van Flandern. On Earth there is a growing body of evidence suggesting the poles have shifted position over time. In 1958, Charles Hapgood proposed a theory to explain the ice ages, which involved sudden shifts or displacements of the Earth's crust over the mantle. Albert Einstein, a supporter of Hapgood's theory summarizes it as follows [8]:

> In a polar region there is continual deposition of ice, which is not symmetrically distributed about the pole. The Earth's rotation acts on these unsymmetrically deposited masses, and produces centrifugal momentum that is transmitted to the rigid crust of the Earth. The constantly increasing centrifugal momentum produced this way will, when it reaches a certain point, produce a movement of the Earth's crust over the rest of the Earth's body, and this will displace the polar regions toward the equator.

There is evidence indicating that during the last ice age, the north pole was located at around 60° N, 83° W, near the middle of Hudson Bay. North America was, at this time, nearer the pole and covered by a massive layer of ice and snow. Hapgood believes the force resulting from this build up was great enough to cause the North American continent to slip 3000 km south toward the equator. On the other side of the world, Siberia, which had a temperate climate prior to the slip, shifted by the same amount northward toward the pole, flash freezing its inhabitants including the mammoths.

Archaeologists have long wondered why certain sacred sites in Central America are aligned as they are. For example, why is the major axis of Teotihuacan, which is oriented 15.42° east of north, not oriented due north? Teotihuacan is generally thought to have been built around the first century A.D. [9]. Rand Flem-Ath believes it is much older. He argues that prior to the last crustal displacement around 9600 B.C., Teotihuacan's Street of the Dead was oriented north-south. When the crust shifted, features west of 83° W, between 83° W and 107° E on the Earth's surface rotated clockwise while those in the opposite hemisphere rotated counterclockwise. Before 9600 B.C., Teotihuacan would have been located near 45° N and would be aligned to within a few degrees of due north given the previous pole position of 60° N, 83° W. Of course this position is only a estimate. If the pole was at a slightly different position, Teotihuacan might have been aligned exactly to the meridian.

Peter Schultz has found compelling evidence that the location of the Martian poles too have changed in the history of the planet [10]. By analyzing grazing meteor impacts, the ages and locations of layered deposits in the equatorial regions, and features indicative of polar or near-polar locations, he has estimated that prior to its present position, Mars' north pole was located at around 45° N, 160° W, northwest of Olympus Mons. Moving further back in time, Schultz shows the path of the pole moves toward equatorial regions. Several billion years ago he believes it was located south of Olympus Mons, near the Tharsis volcanoes.

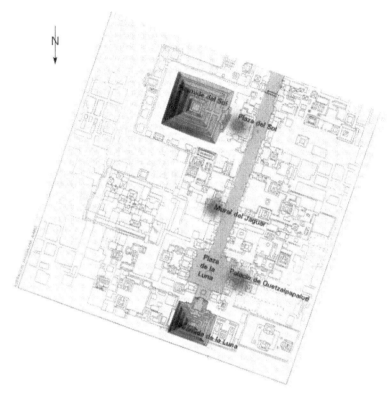

Figure 77 Map of Teotihuacan near Mexico City. Like other sacred sites in middle America it is not, at present, aligned to the cardinal directions. However, it can be shown that prior to the last crustal shift, which is thought to have occurred around 9600 B.C., Teotihuacan, whose principal axis is now 15.42° east of north, would have been oriented north-south.

Since Cydonia is east of 160° W, between 160° W and 340° W, features would have rotated counterclockwise as the pole moved from 45° N, 160° W to its present location. Van Flandern argues that the Face, which is now located at 41° N, 9.5° W and oriented with its major axis 30.9° west of north, would have been located almost exactly on the Martian equator[28]. Given the previous pole position it would have rotated 20.7° clockwise, not enough to assume an exact north-south orientation. But again, Schultz's estimate is only an estimate. If the pole were instead located at 39° N 148° W, a difference in position of about 600

---

[28] http://metaresearch.org/solar%20system/cydonia/mrb_cydonia/new-evidence.asp

km, the Face would have been within a few degrees of the equator and oriented north-south.

Analysis of layered deposits and pedestal craters at the poles suggest the polar regions are relatively young, between 1-100 million years old. If Van Flandern's hypothesis is correct, sometime within the last 100 million years, the City, Face, and other sites in Cydonia would have been aligned with the compass directions. If this is true and these objects are artificial, they are incredibly ancient by contemporary standards.

The last explanation for the alignment is the original one suggested by Hoagland — that it has some kind of archaeo-astronomical significance. Instead of using the angle of Hoagland's line from the City Square through the mouth of the Face (23.5° north of east), if we use the average of the orientations of the Face, Fort, Starfish Pyramid, rounded formation, and mounds (33.3° north of east) we get not one but eight dates within the last half-million years satisfying the summer solstice alignment: 33,000, 120,000, 160,000, 200,000, 260,000, 330,000, 390,000, and 450,000 years B.P. (Figure 73).

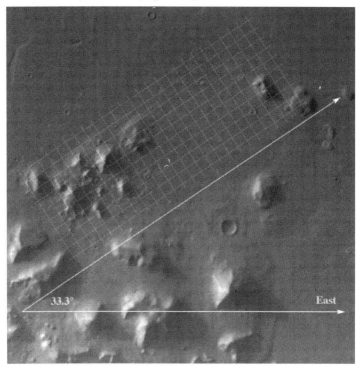

Figure 78 Map projected Viking image of the Cydonia site on Mars. The cross-hatched pattern over the region containing the City and Face is oriented at 33.3° north of east. The southern face of the D&M Pyramid is oriented due south. Van Flandern has shown that the City and Face might have once been located near the Martian equator and oriented to the meridian.

If we compare these dates with significant events that have occurred on Earth during this period there are several interesting coincidences, which we shall explore in the next section. Perhaps, like Stonehenge, the alignment of the Cydonia complex with the summer solstice sunrise indicates its date of construction, or perhaps it marks some other important epoch in the history of Mars, or even Earth.

Terrestrial Connections

The City and Face on Mars remind most people of the pyramids and Sphinx on the Giza plateau in Egypt. Egypt is a good place to begin, first to better understand the scale of the Martian anomalies in comparison to similar features on Earth, second to gain an appreciation for the problem of differentiating eroded architecture from geology, and third to begin to identify possible connections between Earth and Mars.

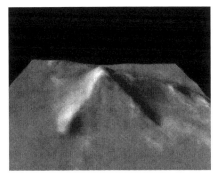 

Figure 79 Perspective view of south side of the D&M Pyramid generated by projecting Viking image 35A72 onto a terrain elevation model (left). The pyramids of Khufu, Khafre, and Menkaure in Giza (right).

The area in the Giza plateau containing the three largest pyramids of Khufu, Khafre, and Menkaure is several square kilometers in size. On Mars, the Face and surrounding objects occupy an area many times larger. Where the distance between the Great Pyramid and the Sphinx is less than a kilometer, the distance between the Face on Mars and the D&M Pyramid is over 20 kilometers. The Great Pyramid of Khufu has four sides. Its height is 135 meters and the length of its base is 229 meters. This wonder of the ancient world is dwarfed by the objects on Mars. The height of the D&M Pyramid is about 1250 meters and the lengths of its sides are between 2700 and 3800 meters — about ten times longer than those of the Great Pyramid. Thus, its area is more than 100 times larger, and its volume over 1000 times greater than that of one of the largest structures on Earth!

Egypt presents us with a variety ancient monuments in various stages of decay. Some are so eroded that they are almost indistinguishable from geological features. Examples include the southern brick pyramid of Amenemhet III as viewed from the north, the pyramids of Userkaf, Pepy I, and Unas in Saqqara, the pyramids of Neferirkare and Neuserre in Abusir as viewed from the southwest, and the Mastaba of Ptahshepses and Sun Temple of Userkaf, both in Abusir. On the other hand there are some geological features such as the rock formations in El-Faiyum that look as if they could be ancient monuments. That it can be difficult in some cases using aerial imagery to distinguish between manmade structures and natural landforms on Earth suggests the potential difficulty in determining the nature of features on another planet such as Mars without actual field work.

Figure 80 Southern brick pyramid of Amenemhet III (left) built in the Middle Kingdom, 12th Dynasty (1979-1801 B.C.) Although this and other similar pyramids might have looked impressive when new, lacking the support of solid stone, once the limestone casing broke, the pyramids collapsed spilling out debris not unlike that seen at the base of eroded natural formations in the American southwest (right).

The Sphinx is one of many structures in Egypt that has undergone extensive repair over the years. Motivated by the work of maverick Egyptologist John Anthony West, geologist Robert Schoch has demonstrated that the patterns of erosion seen on the body of the Sphinx are those of water [11]. This means that the Sphinx probably dates back not to Khafre and the fourth dynasty as most Egyptologists believe, but much earlier, perhaps as far back as the end of the last Ice Age around 9600 B.C. when rainfall was abundant on the Giza Plateau. In that epoch, the Sphinx, a creature with the head of a man and the body of a lion, would have faced the constellation Leo rising due east on the morning of the spring equinox.

Construction engineer and Egyptologist Robert Bauval has identified two interesting connections between the Sphinx and Mars [12]. In certain tombs, Mars is referred to as 'the eastern star' and as 'His name is Horakhti'. The Sphinx, which faces east is also called Horakhti. It is thus not unreasonable to conclude that the Sphinx might, by association, have been called 'Mars'. Another name the ancient Egyptians used for Mars was 'Horus the Red'. Bauval has also determined that for a long period of its history, the Sphinx was painted red.

Other connections between Earth and Mars can be found in ancient Mesopotamia. The Babylonian name for Mars was Nergal, the Sumerian god of the underworld. An interesting connection between Mars, Nergal and the great Flood is found in the *Epic of Gilgamesh*. In it, Utnapishtim, the biblical Noah, tells Gilgamesh how the Flood began [13]:

> With the first glow of dawn,
>
> A black cloud rose up from the horizon.
>
> Inside it Adad [god of Storm and Rain] thunders,
>
> While Shallat and Hanish [Heralds of Adad] go in front,
>
> Moving as heralds over hill and plain.
>
> Erragal tears out the posts [of the world dam];
>
> Forth comes Ninurta [son of Enlil] and causes the dikes to follow.

The tale continues:

> The Annunaki lift up the torches,
>
> Setting the land ablaze with their glare,
>
> Consternation over Adad reaches to the heavens,
>
> Turning to blackness all that had been light.
>
> The wide land was shattered like a pot!
>
> For one day the south-storm blew,
>
> Gathering speed as it blew, submerging the mountains,
>
> Overtaking the people like a battle.

The text states that Erragal, another name for Nergal, and Ninurta release a wall of water, a tidal wave, that smashes into the land. The amount of water released is so great that it covers the mountains. Then comes the storm:

> Six days and six nights
>
> Blows the flood wind, as the south-storm sweeps the land.

Geologist Robert Schoch believes the Flood was the result of an asteroid impact. For the sake of argument he assumes a rocky asteroid roughly 6 miles in diameter colliding with the Earth at a speed of 55,00 miles per hour with the explosive yield of about 1 billion megatons. Such an object impacting in the deep ocean would displace an immense volume of water and create an enormous tidal wave or tsunami. After passing through the water, the force of the impact would trigger severe earthquakes resulting in additional tsunamis.

Puncturing the Earth's crust in the process, hot magma released from the underlying mantle would vaporize huge quantities of water. From the thick clouds formed from the condensing water vapor, torrential rains would fall. The signature of an ocean impact — a tidal wave followed by torrential rains — matches the pattern of events as told in the *Epic of Gilgamesh*.

Zecharia Sitchin contends that the Flood was a natural disaster triggered by the passage of Nibiru through the inner solar system past Earth, one that Enlil used to his advantage in an attempt to eradicate mankind [14]. The *Book of Genesis* tells us of God's displeasure with Man [15]:

> And God looked upon the Earth, and, behold, it was corrupt; for all flesh had corrupted his way upon the Earth. And God said unto Noah, The end of all flesh is come before me; for the Earth is filled with violence through them; and, behold, I will destroy them with the Earth... And, behold, I, even I, do bring a flood of waters upon the Earth, to destroy all flesh, wherein is the breath of life, from under heaven; and every thing that is in the Earth shall die.

This passage suggests the Flood was the result of a deliberate act, one that according to the *Epic of Gilgamesh* was initiated by Nergal, the god of the underworld, who is associated with Mars.

As we have seen, Mars too is planet that has been ravaged by massive floods. Hancock and Bauval contend the Egyptians built the pyramids and Sphinx in Giza to commemorate the 'first time' or Zep Tepi. Is it a coincidence that this event occurred at the end of the last Ice Age following what appears to be a massive world-wide flood? And isn't it strange that the Sphinx, a key element of this memorial, represents the same god who according to Sumerian legend "tears out the posts" of the world dam and starts the Flood?

The Face on Mars has often been compared with the Sphinx. But an even more striking similarity can be found half way around the world, high up in the Peruvian Andes. Just outside of Lima on the Marcahuasi Plateau (11° 46' 40.9" S, 76° 35' 26.3" W) lies an unusual assortment of ancient stone 'sculptures' of animal and human figures. Along the edge of the plateau is a fully rendered 3-D representation of a humanoid head staring up at the sky. Of all the features on Earth, this one is most like the Face on Mars. Although not as large, this face on the Marcahuasi plateau is equally enigmatic.

According to Daniel Ruzo, who first investigated the area in the 1950's, the sculptures "range from natural rock, the form of which has been roughly cut to show features, to the mass of rock perfectly cut to show several persons who can be distinguished by looking from different angles..." [16]. Ruzo, who spent months on the plateau studying these objects, found that many of them exhibited an unusual four-dimensional characteristic changing appearance with the time of day, season, or vantage point. In the face, which Ruzo calls the Inca's Head, he claims that one can see three races of humanity as the lighting changes.

Figure 81 Ninety-two foot high body and face staring skyward on the Marcahuasi Plateau. According to Ruzo, three races of humanity can be seen as lighting conditions change during the day. "It has a sphinx-like look toward infinity...which contemplates man's past on Earth with the eyes of all races." (Bill Cote)

The nature of these objects on the Marcahuasi plateau in Peru is controversial. Ruzo is convinced of their artificiality:

> The number of artifacts in a small area of 3 sq. km. is enormous, several hundred. Nowhere else is there to be seen such a density of 'natural curiosities' in such perfect condition.
>
> Certain shapes, such as human chins and mouths are repeated with such insistence that the hazard of erosion cannot be said to be the cause, except by the faintest of probabilities. This is especially the case with the rock: The Inca's Head.
>
> The points from which the figures can be observed are indicated by markers, some of which permit [one] to see several figures over an arc of 90 degrees. Such a coincidence would appear to be impossible.

According to Ruzo, "traditions speak of the giants or Huaris, legendary beings, builders of the cyclopean structures, the remains of which persisted until the Incas." Film producer Bill Cote who visited the Marcahuasi Plateau in 1989 had this to say[29]:

> What little is written about Marcahuasi indicates a certain reluctance on the part of archaeologists to say that the figures are man-made. Indeed, many of them are subtle and not always obvious to the viewer. But that is precisely what contributes to the mystery. There are so many recognizable forms there, that one is tempted to say they must be man-made, or else nature is having a great joke on us.

Objects so old that mainstream scientists refuse to acknowledge their presence. Two faces on neighboring planets staring at each across the vastness of space. Is

---

[29] http://www.science-frontiers.com/sf078/sf078a03.htm

nature playing a joke on us as mainstream scientists would like us to believe, or are all of these 'coincidences' trying to tell us something very important?

 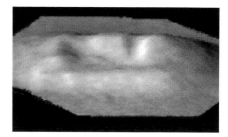

Figure 82 Two faces staring at each other acoss the vastness of space (Cote/Carlotto)

Return to Myth

We have traced the search for extraterrestrial life, from its roots in myth through to the Cydonia discoveries and the latest scientific evidence gathered by Mars Global Surveyor and other probes. In seeking terrestrial connections between Earth and Mars, we return to the beginning, to myth, and to those age-old questions: Who are we, and where did we come from?

Although ancient texts suggest answers to these questions — answers that, to Zecharia Sitchin and others, indicate the intervention of extraterrestrials in human affairs — because of the lack of physical evidence, scientists have been unwilling to seriously consider the extraterrestrial possibility. If, however, these structures on Mars are artificial, that can all change.

The philosopher Karl Popper proposed that the best way of evaluating a scientific hypothesis is to determine whether or not it can be proven false. By definition, a hypothesis is falsifiable if there exists an observation that is logically inconsistent with it. Strictly speaking, radio SETI is not falsifiable because there are always more directions to look and other ways to listen in for extraterrestrial radio signals. Unlike radio SETI, which has searched for more than a half-century in vain for extraterrestrial signals, the question of whether there was once intelligent life on Mars in Cydonia is falsifiable, and because it is falsifiable, it is a legitimate scientific hypothesis. Perhaps we will not know for sure until that day when archaeologists dig in the red sands of Mars, but it is a question that can eventually be answered one way or the other.

If one day the Face and other objects are found to be natural, then this, the latest chapter in the search for extraterrestrial intelligence will come to a close. It will be a disappointment, but from a history of science perspective important new lessons will have been learned.

But what if we find that even one of these objects on Mars is artificial? Hoagland originally suggested three possibilities: that they were built either by native Martians, or a technologically-advanced, prehistoric-race from Earth, or extraterrestrials. Any one of these possibilities leads to extraordinary implications for science and humanity.

Darwin's theory of evolution holds that where conditions are suitable, life will develop and, over a very long period of time, evolve into more complex organisms. If the Face was built by native Martians, then intelligent life was able to develop on Mars over a much shorter period of time than on Earth, either during the first billion years or so while Mars possessed a thick atmosphere and liquid water, or during one of its brief episodic periods of climate change. This would represent a serious challenge to classical evolutionary theory.

Because evolution is, by definition, a random process, the probability that the same life form would evolve in two different places is small. If we discover the humanoid form developed independently on Mars, Darwinian evolution will be dealt another blow. Perhaps then the humanoid form is not the result of terrestrial evolution, but is something else.

The origin of life on Earth is a mystery. Some scientists believe the Earth has been 'seeded' by extraterrestrial microorganisms carried by meteorites, an idea known as 'panspermia'. Although panspermia may explain the appearance of the earliest primitive life forms on Earth, it cannot explain the appearance of advanced life forms that formed during the Cambrian explosion, or the sudden appearance of Man in the last several hundred thousand years.

Could advanced life have been introduced to Earth from somewhere else? Sitchin claims the Annunaki came from a yet to be discovered planet, Nibiru, well beyond the orbit of Pluto. But could they have come from Mars instead?

It is written in the Book of Genesis that:

> In the day that God created man, in the likeness of God made he him;
>
> Male and female created he them; and blessed them, and called their name Adam.

According to Sitchin, Adam was derived from the Hebrew 'adama' which means dark red soil. Moreover 'adama' and the Hebrew name for the color red 'adom' stem from the words for blood, 'adamu' and 'dam'. According to Sumerian legend, Man was created by Enki and Ninhursag in the Underworld, which Sitchin believes was southern Africa. This location is consistent with the analysis of mitochondrial DNA that indicates Man (Homo sapiens sapiens) first appeared in southern Africa roughly 200,000 years ago. But consider several more coincidences:

1. Nergal, the god associated with Mars, was Lord of the Underworld.
2. Mars is red because of its oxidized iron-rich soil, which incidentally, is why blood is red.
3. The time period Man first appeared in southern Africa (200,000 B.P.) satisfies one of the solstice alignment of the Cydonia complex — it is

one of eight epochs within the last half million years when the sun rose in line with the City and Face on the first day of summer.

Recall the Cydonia complex also satisfies a solstice alignment 445,000 years B.P., around the time Sitchin claims the Annunaki first came to Earth. Could the alignment of the Cydonia complex be an elegant way of marking these two important epochs?

Another possibility is that instead of coming from Mars to Earth, humans went from Earth to Mars. Although most scholars believe that writing originated in Sumeria in the fourth century B.C., there is evidence that it is much older [17]. Etymological studies point to a fully modern human language in existence as long as 40,000 years ago and writing as far back as 15,000 B.C. Among the eight solsticial alignments for the Cydonia complex over the last half-million years, the last one occurred about 33,000 years B.P. Could there have been contact between Earth and Mars then, or at some other time in the distant past? Are these objects on Mars the ruins of a previous age of Man?

The third possibility is that the Face was built by extraterrestrial visitors to our solar system. Carl Sagan estimated that such contact might occur every 10,000 years or so. If this is true, it seems likely that we will eventually discover other extraterrestrial artifacts within our solar system. Perhaps the Face on Mars is the first.

As unbelievable as these possibilities may seem, if there are artificial structures on Mars, then one of them has to be true. An alternative to classical evolutionary theory may have to be found. Human history will certainly have to be rewritten. We may learn that we are not alone, that humanity is far older than we think, or both.

Sources

1. Michael J. Crowe, *The Extraterrestrial Life Debate, 1750-1900*, Dover Publications, Mineola NY, 1999.
2. S. V. McDaniel, *The McDaniel Report*, North Atlantic Books, Berkeley CA, 1994.
3. Steven J. Dick, *Plurality of Worlds, The Origins of the Extraterrestrial Life Debate from Democritus to Kant*, Cambridge University Press, Cambridge, 1981.
4. Kevin Greene, *Archaeology: An Introduction,*" University of Pennsylvania Press, Philadelphia, 1983.
5. W. Ward, "Large-scale Variations in the Obliquity of Mars," *Science*, Vol. 181, pp 260-262, July 1973.
6. Mark Carlotto, "Enigmatic Landforms in Cydonia: Geospatial Anisotropies, Bilateral Symmetries, and Their Correlations," *Sixth International Conference on Mars*, Pasadena, CA, July 20-25, 2003.

7. J.W. Head, H. Hiesinger, M.A. Ivanov, M.A. Kreslavsky, S. Pratt, and B.J. Thomson, "Possible Ancient Oceans on Mars: Evidence from Mars Orbiter Laser Altimeter Data," *Science*, Vol. 286, pp 2134-2137, 10 December 1999.

8. Rand and Rose Flem-Ath, *When the Sky Fell: In Search of Atlantis*, St. Martin's Press, 1995.

9. E.C. Krupp (ed.), *In Search of Ancient Astronomies*, McGraw-Hill, New York, 1978.

10. P.H. Schultz, "Polar Wandering on Mars," *Scientific American*, Vol. 253, pp 94-102, 1985.

11. Robert M. Schoch, *Voices of the Rocks*, Harmony Books, New York, 1999.

12. Graham Hancock and Robert Bauval, *The Message of the Sphinx*, Crown Publishers, New York, 1996.

13. "Myths Of The Flood: The Flood Narrative From The Gilgamesh Epic," Translation by E. A. Speiser, in *Ancient Near Eastern Texts*, Princeton, 1950, pp. 60-72.

14. Zecharia Sitchin, *The 12th Planet*, Avon Books, New York, 1976.

15. *King James Bible* (http://quod.lib.umich.edu/k/kjv/).

16. Daniel Ruzo, "The Masma Culture," *L'Ethnographic*, Paris, 1956.

17. Richard Rudgley, *The Lost Civilizations of the Stone Age*, The Free Press, New York, 1999.

Figure 83 The Flood Tablet, the eleventh tablet from the *Epic of Gilgamesh* recovered from the ruins of the ancient library of Nineveh in northern Iraq. (Courtesy Trustees of the British Museum)

Eleven — Epilog

> Apart from the question of whether the landforms are the product of natural or intelligent process, the question of other intelligent life forms in the universe raises questions about the nature of consciousness... — Randy Pozos

After more than half a century of listening, SETI has not heard an extraterrestrial radio signal. Thirty years after Viking was sent to Mars to look for evidence of life, a live microbe has yet to be found. Even after extensive high-resolution imaging of the Face and other Martian anomalies, no obvious indication of artificiality has been detected — no roads or buildings.

One has to wonder if the answer to the question of whether or not there is other life in the universe is beyond our grasp. Are we in some kind of a 'cosmic quarantine' as James Deardorff has suggested? Perhaps, like in *Star Trek*, no one will contact us until we prove ourselves by achieving an advanced level of technology, or a higher state of consciousness.

Another possibility is that we have simply been looking for the wrong things, and perhaps making the wrong assumptions.

A Signature of Artificiality

As we saw earlier, imaged close-up, ancient ruins on Earth are often indistinguishable from geology. Depending on their age and degree of degradation, artificial constructions break down and begin to blend into the natural background. It is simply the 2nd Law of Thermodynamics, the universal tendency toward disorder, at work. The gradual breakdown of a structure can be measured by the Fourier power spectrum of an image of the structure. Most people are familiar with the concept of sound spectrum; e.g., a bass guitar has most of its energy in low frequencies, a flute in high frequencies, and the human voice in between. Sound emanating from a loudspeaker is a one-dimensional signal (pressure wave). Images are two-dimensional and so have a 2-D power spectrum. In addition to frequency, 2-D power spectra have a directionality. As seen in Figure 84, a city has a highly directional power spectrum that spans a range of frequencies. The directionality is caused by the alignment of streets and buildings in a rectangular grid. As artificial structures degrade over time — buildings collapse, streets become filled-in with debris, etc. — the higher spatial frequencies (those farther away from the center) fade away. We see this in La Centinela, where the extent of the power spectrum is significantly reduced. Yet a strong directional component remains. Extreme degradation in Sipan further reduces the higher spatial frequencies to the point where little or no directionality remains.

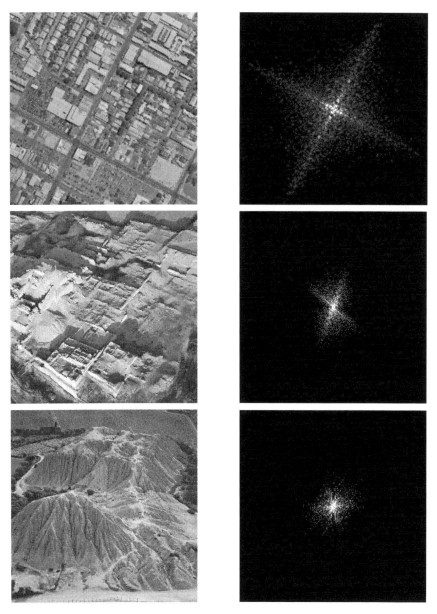

Figure 84 Three images (left) and their Fourier power spectrum (right). The three images are (top to bottom): a city in the US, La Centinela, and Sipan, both in Peru. La Centinela was built several hundred years before the Inca occupation in 1450. The eroded pyramids in Sipan have been dated to around 300 A.D. (Images of Peru courtesy Marilyn Bridges)

What is interesting in images over Cydonia is that they have a directional 'signature' that is similar to highly eroded terrestrial archaeological ruins (Figure 85). In a study of the geospatial statistics of imagery over Cydonia entitled "Enigmatic Landforms in Cydonia: Geospatial Anisotropies, Bilateral

Symmetries, and Their Correlations" presented at the *Sixth International Conference on Mars* in 2003, I showed that these directions line up with the symmetry axes of objects in the City (Figure 28), including the Face, Fort, and Starfish Pyramid, and with the D&M Pyramid (Figure 72). Even though most of the scientists at the conference would probably have laughed at the idea of the Face on Mars being artificial, none offered an explanation for this alignment, other than the crustal dichotomy, which itself is unexplained.

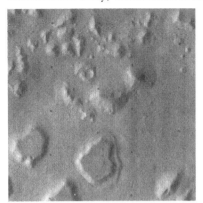 

Figure 85 Viking orbiter image of a rectilinear formation southwest of the City (left) and its Fourier power spectrum (right). Note similarity to those in previous figure.

Patterns of a Technological Intelligence

Can this 'signature of artificiality' be quantified so that we can compare patterns on Mars (and elsewhere) with terrestrial images? In 1987, Michael Stein developed a technique for detecting manmade objects in images based on fractal models. The method assumes that natural backgrounds are fractal (within a certain range of scale) and manmade objects are not. Under certain conditions, the detector generates false alarms — natural features that are not fractal. One example is cast shadows at low sun angles.

The fractal object detector uses the fractal dimension, which measures the 'roughness' of the image, and model fit error (how well the image fits a fractal model). It occurred to me that by combining the fractal measurements with directional measurements from the power spectrum one might be able to do a better job of differentiating between natural and artificial patterns. More important than the actual direction in the power spectrum is the degree of directionality, or anisotropy. Another important feature is the degree of rectilinearity — whether there are directional components 90° apart.

Consider two sets of images, one containing natural backgrounds and textures (Figure 86), the other, manmade objects and structures (Figure 87).

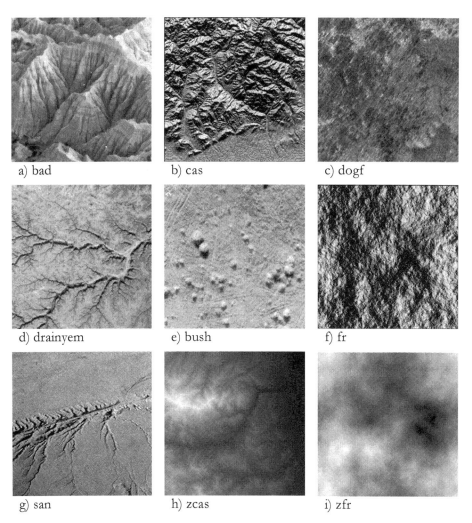

Figure 86 Aerial images of natural terrestrial (and synthetic fractal) backgrounds.

a) Badlands in western South Dakota (US). The Badlands contain a variety of formations resulting from severe erosion of the clay-rich soil by wind and rain.
b) Mountains near the Caspian Sea. The image is a computer graphics rendering (shaded rendition) of the elevation map in h).
c) Forested area in winter on Cape Ann Massachusetts (US). Cast shadows from pine trees produce a strong directional component in the image.
d) Image of a drainage pattern in Yemen. (Similar drainage patterns can be found on Mars.)
e) Aerial image containing trees and bushes. Some vehicle tracks are also evident on the ground.
f) Computer graphics rendering of the synthetic fractal surface in i).
g) Section of the San Andreas fault in California (US).
h) Elevation map of mountains near the Caspian Sea.
i) Synthetic fractal surface (D=2.5).

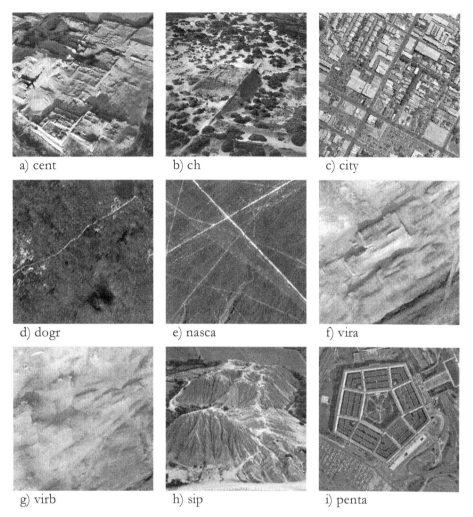

a) cent　　　　　　b) ch　　　　　　　c) city

d) dogr　　　　　　e) nasca　　　　　　f) vira

g) virb　　　　　　h) sip　　　　　　　i) penta

Figure 87 Aerial images of manmade objects and features. (Images of Peru courtesy Marilyn Bridges [2].)

a) La Centinela (Peru). These structures were built several hundred years before the Inca occupation in 1450.

b) Chotuna (Peru). Adobe pyramid and surrounding irrigation canals were built around 600 A.D.

c) Urban area in the US.

d) Abandoned road on Cape Ann Massachusetts. The road, which was built by colonial settlers, has not been used since the early 1800's.

e) Nazca lines, Peru. The origin and purpose of this vast network of lines and geoglyphs is unknown.

f) Structure within unexplored ruins in Viru Valley (Peru).

g) Less distinct section of ruins in Viru Valley.

h) Sipan (Peru). Eroded pyramids dated to around 300 A.D.

i) Pentagon building (US). j) Military tank. k) Military truck.

j) tank

k) truck

How well can we distinguish the images in these two sets using the four measurements mentioned above, namely the fractal dimension, model fit, anisotropy, and rectilinearity? To make it fair (and challenging) notice some of the images of natural patterns have strong linear features while some of artificial patterns have very little linear structure. It turns out that 85% of the images can be correctly classified (Figure 88). Three mistakes were made: 'vira' was classified as a natural pattern, and 'zcas' and 'zfr' were both classified as artificial patterns.

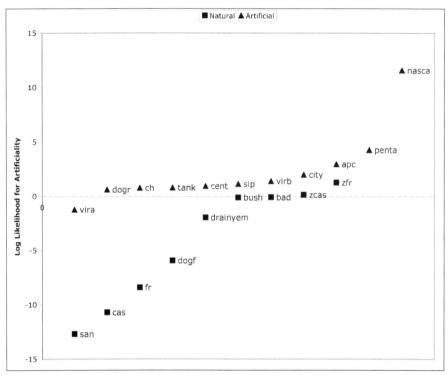

Figure 88 Plot of log likelihood values of artificiality for terrestrial patterns. Artificiality indicated by values greater than zero (y-axis).

The results of applying this classifier that was 'trained' on terrestrial imagery, to a third set of images of 'enigmatic' features on Mars, the Moon, and elsewhere (Figure 89) are intriguing [3]. Areas on the Moon exhibiting rectilinear characteristics have high scores for artificiality; so do areas on Mars, including the area in Figure 85. The Face on Mars is on the edge, with a value close to zero. Values neither strongly positive (artificial) nor strongly negative (natural) are indeterminate. Scores for eroded archaeological ruins on Earth are in this range.

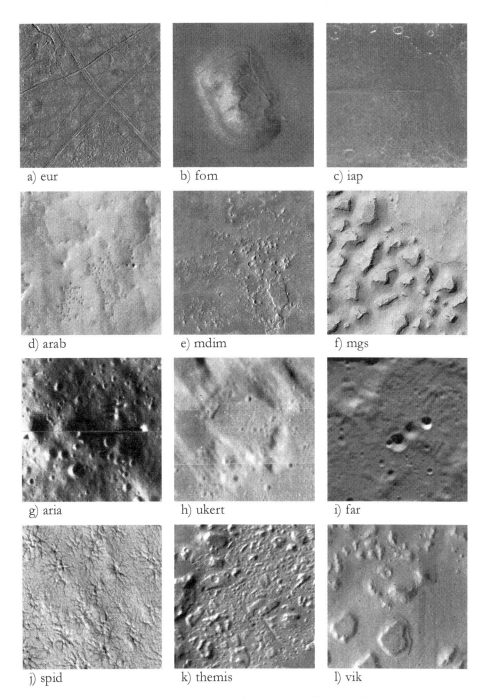

Figure 89 Satellite images of 'interesting' features from Europa, Mars, Iapetus, and the Moon (NASA).

(continued)
a) Ice flows on Jupiter's moon Europa (Galileo).
b) "Face on Mars" (Mars Global Surveyor).
c) Rim around Saturn's moon Iapetus. Image from a section of an image map constructed from a number of images taken by the Cassini spacecraft.
d) Portion of Arabia Terra on Mars containing a number of tracks lefts by dust devils (MGS).
e) Section of Mars Digital Image Map over Cydonia region of Mars. Derived from Viking orbiter imagery.
f) Mesas on the Elysium Plains of Mars (MGS).
g) Lunar Orbiter image of rectangular depressions near the crater Ariadaeus B on the moon.
h) Lunar Orbiter image of a rectilinear formation next to the crater Ukert.
i) Clementine image of rectilinear "scrapings" on the far side of the moon.
j) Unusual formations ("spiders") around Chasma Australe at the south pole of Mars [4].
k) THEMIS image of rectilinear texture in Cydonia, Mars.
l) Rectangular arrangements of mesas in Cydonia (Viking Orbiter).

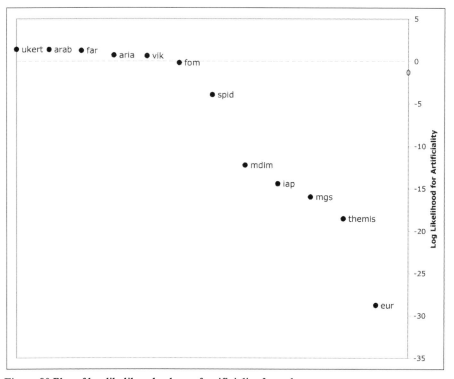

Figure 90 Plot of log likelihood values of artificiality for unknown patterns.

Each pattern has a set of four measurements. Using these values, we can compute the 'distance' between patterns. Two patterns that have similar measurement values are 'close' to one another. Table 4 lists terrestrial analogs of some of the more enigmatic features. The rectilinear area around the crater Ukert on the Moon is similar to ruins in the Viru Valley in Peru. The region in

Cydonia imaged by Viking is similar to ruins in a different area in the Viru Valley. The MGS image of the Face on Mars has similarities to both natural and artificial textures — two are highly eroded archaeological structures, and one is a drainage pattern.

Table 4 Distances between top six enigmatic patterns and their nearest terrestrial analogs

ukert	vik	arab	aria	fom	far
virb (1.19)	vira (0.97)	bush (0.99)	zfr (0.32)	bush (0.21)	zfr (1.09)
		dogr (1.06)	bad (0.53)	cent (0.45)	virb (2.26)
			virb (0.67)	drainyem (0.59)	
				ch (1.07)	

Although the data are too limited in size and scope to draw any definite conclusions, the results suggest that certain areas on Mars and the Moon appear artificial by comparison with terrestrial features.

Reconsidering Basic Assumptions

Enigmatic features on Mars that resemble ancient ruins on Earth, occurring within a geological context that cannot itself be explained. Most scientists dismiss the very idea of advanced life on Mars because it conflicts with the theory of evolution, which assumes that incredibly long periods of time are required for advanced life to evolve from more primitive forms.

The theory of evolution is thought by most to be the only scientific explanation for the diversity of life on Earth. But is it?

Proposed more than one hundred and fifty years ago, the essence of Darwin's theory of evolution is that over time changes in the environment lead to changes within a species [5]:

> Now, if nature had to make the beak of a full-grown pigeon very short for the bird's own advantage, the process of modification would be very slow, and there would be simultaneously the most rigorous selection of the young birds within the egg, which had the most powerful and hardest beaks, for all with weak beaks would inevitably perish...

Darwin believed that over even longer periods of time, the process of natural selection led to the evolution of new species:

> I am fully convinced that species are not immutable; but that those belonging to what are called the same genera are lineal descendants of some other and generally

extinct species, in the same manner as the acknowledged varieties of any one species are the descendants of that species. Furthermore, I am convinced that Natural Selection has been the main but not exclusive means of modification.

This, according to molecular biologist, Periannan Senapathy, is the fatal flaw of Darwin's theory. Published a century before the discovery of DNA, Darwin did not know about the genetic basis of life — that the genome, which is composed of long sequences of DNA, much like a computer program, is what determines the organism. Senapathy has shown that the time it would take random genetic mutations in nature to change a functional sequence of DNA, known as a gene, into another, is something like 10^{350} years. He argues from a statistical standpoint that "The generation of variants of the same gene is very easy, but an entirely new gene is simply improbable, even over geological time" [6].

Although one might think simple organisms have simple genomes, it turns out that the complexity of an organism is not indicative of the size of its genome. In fact, some organisms have genomes that are even larger the human genome. For example, a kind of plant known as *Fritillaria assyrica* has 1.3×10^{11} base pairs on its genome, while the human genome is 100 times smaller, having only 3.2×10^9 base pairs[30].

In his book, *Independent Birth of Organisms*, Senapathy describes a new theory to explain the diversity of life on Earth. He assumes the 'primordial pond' — the pond where life was supposed to have started — was rich in DNA, so rich that the probability of it randomly organizing into functional DNA sequences (genes) was high. He further assumes that genes from this 'pool' of genes could organize, in turn, into higher-level genetic structures and ultimately into genomes. The genome is the blueprint of the organism, containing instructions that define the developmental-genetic pathway of the organism — the biological program that controls how the organism forms and develops. He believes that of all the organisms that could form, only those that can survive in a given environment will form. According to his theory, there are no 'missing links' because organisms do not evolve from one form (species) into another. Instead, when conditions are right, many varieties form independently, at around the same time, like they did during the Cambrian Explosion. In contrast with Darwin's view that "that species are not immutable", Senapathy instead argues:

> Each independent-born organism has a constant set of genes in its genome, and the DG [developmental genetic] pathway of each independent creature is rigid. Therefore, the independently-born organism is immutable [6].

Where Darwin saw the lack of transitional forms as a problem with the fossil record — that it was incomplete, Senapathy sees it as a direct consequence of independent births.

---

[30] Not all of the DNA forms useful genes. Much of it, called 'junk DNA', does not seem to have any function at all.

Ironically, for all of its scientific appeal, Darwin's theory lacks any explanation for how life began:

> There is grandeur in this view of life, with its several powers, having been originally breathed into a few forms or into one; and that, whilst this planet has gone cycling on according to the fixed law of gravity, from so simple a beginning endless forms most beautiful and most wonderful have been, and are being, evolved [5].

Senapathy's theory picks up from the Chemical Evolution Hypothesis leaves off, and proposes the first truly scientific (i.e., testable) explanation for how life might have formed from early Earth's chemical 'soup'.

Returning to Mars, the implication of Senapathy's *Independent Birth of Organisms* hypothesis is clear — if conditions on Mars were right in the past, like on Earth several times in its history, then a variety of complex life forms, perhaps even advanced life forms, might have formed in a relatively short period of time. If this is true, then the existence of artifacts from an ancient technological civilization that could have then developed on Mars is not implausible. Perhaps this occurred billions of years ago when conditions on Mars were able to support life. If this was the case, then these structures are incredibly ancient, explaining why they appear as eroded as they do today.

But there is another common assumption — one so fundamental most take it as fact — that may be preventing us from learning the true nature of these objects on Mars, as well as our own origin here on Earth. According to Michael Cremo:

> The time concept of modern archeology, and modern anthropology in general, resembles the general cosmological-historical time concept of Europe's Judeo-Christian culture. Differing from the cyclical cosmological-historical time concepts of the early Greeks in Europe, and the Indians and others in Asia, the Judeo-Christian cosmological-historical time concept is linear and progressive. Modern archeology also shares with Judeo-Christian theology the idea that humans appear after the other major species[31].

In contrast with current linear progressive-time Western scientific paradigms, Eastern spiritual traditions place humanity in the context of repeating time cycles or *yugas*, which last for 24,000 years [7]. Like the story told by Plato in the *Timaeus*, human history is not just one age, the latest age, but a series of ages, punctuated by global catastrophes.

In recent years, what had been the dominant paradigm in geology — that of uniformitarianism, has acknowledged the effect of catastrophic events on Earth's history. Today, it is widely accepted that meteor impacts and other major events have shaped and will continue to shape our planet. However, an

---

[31] http://www.mcremo.com/lectures.html

openness to alternative paradigms in archaeology and anthropology, which are based heavily on Darwinian evolution, has not occurred.

In their book, *The Hidden History of the Human Race*, Cremo and co-author Richard Thompson build a compelling case for the existence of intelligence life on Earth over much of its history. They catalog a variety of artifacts including metallic spheres found in a South African mine dated to 2.8 billion years B.P., a metallic tube found in a 65 million-year-old chalk bed in France, a 200,000- to 400,000-year-old copper coin found in Illinois, and many others. Equally compelling is evidence of modern humans in the fossil record, of human bones as far back as 45 million years ago, and a 2-million-year-old clay figurine found in Idaho.

In the course of their research, Cremo and Thompson show how mainstream scientists have suppressed this information, by saying that it goes against the theory of evolution, and other accepted paradigms — dismissing these anomalies in much the same way NASA has dismissed the Face and the other objects in Cydonia — by saying that because they cannot be there, they are not.

The existence of artificial objects in Earth's strata millions and perhaps billions of years old, offers a new perspective on these mysterious formations on Mars — that the existence of eroded archaeology on Mars (and elsewhere in our solar system) is not impossible, but possible, and perhaps, even probable.

It is interesting how the discovery of possible archaeological ruins on Mars comes at a time when the 'theory' of evolution is beginning to crack, and the truth about Earth's fossil record is finally coming to light. As our planet warms and the oceans rise, we face a potential planetary-scale catastrophe. We are at a turning point and need a paradigm change. Perhaps the Face on Mars is trying to tell us something — something very important.

Sources

1. Mark Carlotto, "Enigmatic Landforms in Cydonia: Geospatial Anisotropies, Bilateral Symmetries, and Their Correlations," *Sixth International Conference on Mars*, Pasadena, CA, July 20-25, 2003.

2. Marilyn Bridges, *Planet Peru, An Aerial Journey through a Timeless Land*, Kodak/Aperture Books, 1991.

3. Mark J. Carlotto, "Detecting Patterns of a Technological Intelligence in Remotely-Sensed Imagery," *J. British Interplanetary Soc.*, Vol. 60, pp 28-39, 2007.

4. Peter K. Ness and Greg M. Orme, "Spider Ravine Models and Plant-Like Features on Mars - Possible Geophysical and Biogeophysical Modes of Origins," *J. Sci. Exploration*, Vol. 55 No 3/4, March-April 2002, pp 85-108.

5. Charles Darwin, *On the Origin Of Species*, 1859 (http://www.gutenberg.org/etext/1228).

6. Periannan Senapathy, *Independent Birth of Organisms*, Genome Press, Madison WI, 1994.

7. Swami Sri Yukteswar, *The Holy Science*, Self-Realization Fellowship, Los Angeles CA, 1990.

8. Michael A. Cremo and Richard L. Thompson, *The Hidden History of the Human Race*, Govardhan Hill Publishing, Badger, CA, 1994.

Index

Abusir, 168
Adam, 6, 173
Africa, 173
Albee, Arden, 4, 99, 111, 115, 122
Aldrin, Buzz, 38
ALH84001, 80, 92, 93, 96
alignment, 32, 49, 51, 52, 160, 164, 166, 167, 173, 174, 177, 179
American Geophysical Union (AGU), 123, 129, 152
anisotropy, 179, 182
Annunaki, 7, 169, 173, 174
Antarctica, 92, 93
Anu, 7, 9
Apkallu, 6
Apollo, 33, 34, 35, 38, 42, 43
Applied Optics, 62, 79, 105, 123
Aquinas, Thomas, 14
Archaeology, vi, 158, 174
Ares Valles, 87
Ariadaeus, 184
Aristarchus, 10, 15, 16
Aristotle, 12, 13, 14, 24
Arizona State University (ASU), 151
Arkhipov, Alexey, 35, 42
Ashurbanipal, 8
Assyrian, 8
Astronomical Diaries, 9
Atlantis, 10, 129, 175
Babylonians, 7, 8, 9, 142
Bacon, Roger, 14
Badlands, 180
Baker, Vic, 89, 90, 96
Bauval, Robert, 168, 170, 175
Berkeley, 56, 58, 79, 80, 95, 112, 174
Berlitz, Charles, 10
Bible, 6, 23, 175
Bowl, 133, 161, 162
Boyce, Joseph, 110
Bradbury, Ray, 81
Brahe, Tycho, 16
Brandenburg, John, 2, 50, 51, 53, 54, 56, 57, 81, 85, 95, 96, 110, 114, 122, 123, 132, 136, 152
Bridges, Marilyn, 95, 178, 181, 189
Brookings Institute, 36, 40
Bruno, Giordano, 15, 18
Buffon, Georges, 21
Burnet, John, 10, 23
Bush, Vannevar, 26
Caltech, ii, 27, 40
Cambrian Explosion, 94, 95, 173, 186
Canals, v, 21
Cantril, Hadley, 27
Carlotto, Mark, i, ii, iv, 42, 79, 80, 95, 103, 104, 105, 106, 132, 172, 174, 189
Carr, Michael, 96, 115, 152
Cassini, Giovanni, 19, 31, 184
catastrophism, 21
Catholic Church, 14, 18, 21
Cattermole, Peter, 82, 95
Channon, James, 2, 51, 71, 136
Chemical Evolution Hypothesis, 93, 187
Christian, 5, 14, 15

Cicero, 5, 10

City, 47, 48, 49, 50, 51, 52, 53, 55, 56, 63, 68, 73, 74, 75, 77, 80, 81, 83, 84, 90, 94, 97, 99, 126, 128, 154, 160, 161, 163, 165, 166, 167, 174, 179

Clarke, Arthur C., 1, 43, 56, 135

Clementine, 34, 184

Cliff, 49, 53, 68, 77, 131, 132

CO_2, 86, 88, 89, 90

Cocconi, Guiseppe, 29

Coffey, Timothy, 100

Condon, Edward, 37, 38, 40, 42

conspiracy, iii, 123, 124

Cooper, Gordon, 38, 42, 95, 129

Copernicus, 16

Cordell, Gene, 50, 51

Cote, Bill, 171, 172

Crater, Horace, 68, 69, 75, 80, 95, 110, 127, 128, 129, 132, 151, 159

Cremo, Michael, 187, 188, 189

Critias, 11

Cros, Charles, 28

Crowe, Michael, 153, 155, 156, 174

Crustal Dichotomy, 81

Cunningham, Glenn, 110, 111, 115, 128

Cusa, Nicholas of, 14

Cuvier, Charles, 21

D&M Pyramid, 46, 51, 65, 66, 67, 68, 76, 101, 104, 114, 125, 126, 128, 150, 151, 159, 161, 162, 166, 167, 179

Darwin, Charles, 21, 173, 185, 186, 187, 189

de Fontenelle, Bernard, 19

Deardorff, James, 30, 42, 177

Democritus, 11, 24, 174

Descartes, Rene, 20

Deuteronilus, 53

Dick, Steven, 11, 14, 24, 157, 174

DiPietro, Vincent, 1, 45, 46, 47, 49, 50, 51, 56, 59, 65, 70, 96, 97, 100, 102, 106, 110, 114, 122, 123

Disney, Walt, 28

DNA, 173, 186

Dolphin, Lambert, 2, 50, 51, 52, 53

Drake, Frank, 29, 30, 155, 156

Drasin, Daniel, 74, 102, 103, 104, 154

Dune spots, 135

EETA 79001, 92

Egypt, 10, 50, 65, 92, 167, 168

Einstein, Albert, 164

Eisenhower, Dwight, 26, 27, 41

Elijah, 6

Endymion, 34

England, Jill, 76, 132, 134, 152

Enki, 7, 173

Enlil, 7, 9, 169, 170

Epicurus, 10, 12, 14, 23

Ereshkigal, 7

Erjavec, James, 76, 80, 83, 84, 85, 95, 128, 132, 136, 152

Erragal, 169

Europa, 183, 184

evolution, iii, 5, 21, 22, 23, 43, 58, 94, 95, 98, 156, 173, 185, 188

Ezekiel, 6

Face on Mars, iii, iv, 1, 2, 3, 4, 44, 45, 46, 47, 48, 49, 50, 51, 52, 53, 54, 55, 56, 57, 58, 59, 60, 61, 62, 63, 64, 65, 68, 70, 71, 72, 73, 74, 75, 76, 77, 78, 79, 81, 83, 84, 85, 90, 94, 95, 97, 99, 101, 102, 103, 104, 105, 106, 107, 109, 111, 112,

113, 114, 115, 116, 117, 118, 119, 120, 121, 122, 123, 124, 127, 128, 129, 130, 131, 133, 136, 137, 139, 140, 141, 142, 147, 148, 149, 150, 151, 152, 153, 154, 156, 157, 158, 160, 161, 162, 163, 165, 166, 167, 170, 172, 173, 174, 177, 179, 182, 184, 185, 188

FACETS, 139

Flammarion, Camille, 22

Fleming, Lan, 115, 140, 152

Flood, 7, 8, 10, 11, 169, 170, 175, 176

Fontana, Francesco, 19

Fort, 48, 73, 74, 75, 77, 126, 130, 161, 166, 179

Fourier, Joseph, 177, 178, 179

fractals, 62, 63, 64, 70, 72, 78, 103, 104, 154, 156, 179, 180, 182

Friedman, Stanton, 26, 41

Galilei, Galileo, 24

Garvin, James, 140

Gassendi, 31

Gauss, Carl, 28, 67, 142

Genesis, 5, 63, 170, 173

Gersten, Peter, 139

Gilgamesh, 7, 169, 170, 175, 176

Giza, 167, 168, 170

glaciers, 84, 88, 90

global warming, 89

Goddard Space Flight Center, 1, 45

Goldin, Daniel, 92, 100, 139

Greenwald, Michael, 55, 56

Grossinger, Richard, 56

Gruithuisen, Franz von Paula, 31, 32, 33, 153

Haas, George, 31, 148

Haisch, Bernie, 123

Haldane, J. B. S., 93

Hancock, Graham, 10, 170, 175

Hapgood, Charles, 164

Head, James, 163

Hebrew, 173

Herodotus, 9

Herschel, Sir William, 2, 20, 49

Hertz, Heinrich, 28

Hipparchus, 10, 15

Hoagland, Richard, 1, 45, 47, 48, 49, 50, 51, 52, 53, 54, 55, 56, 57, 58, 62, 66, 67, 68, 73, 76, 77, 80, 81, 83, 95, 99, 100, 101, 104, 105, 134, 148, 153, 156, 160, 166, 172

Horakhti, 168

Huaris, 171

Huygens, Christiaan, 19, 57

Iapetus, 183, 184

Icarus, 56, 61, 62, 64

ice, 2, 20, 21, 39, 84, 86, 88, 90, 116, 135, 157, 164

Ice Age, 168, 170

image enhancement, 58, 115, 117

Inca, 170, 171, 178, 181

Ishtar, 7

Jakosky, Bruce, 96

Jansky, Carl, 29

Jet Propulsion Laboratory (JPL), ii, iii, iv, 1, 2, 3, 4, 43, 44, 46, 58, 59, 77, 86, 87, 88, 98, 100, 105, 109, 111, 113, 114, 115, 116, 117, 118, 120, 121, 122, 127, 128, 133, 134, 135, 140, 151, 163

Journal of Scientific Exploration (JSE), 123, 124

Journal of the British Interplanetary Society (JBIS), 64, 104

Kant, Immanuel, 20, 21, 24, 174

Kasher, Jack, 39
Kepler, Johannes, 16, 17, 18, 157
Khafre, 167, 168
Khufu, 167
Kiviat, Robert, 112
Kuhn, Thomas, 97, 112
Kynthia, 72
Laverty, David, 102
Leonard, George, 33, 34, 42
Leucippus, 11
Levasseur, J. P., 133
Levin, Gil, 91, 96
little green men, 93
Lowell, Percival, 2, 4, 22, 23, 24, 25, 55, 153, 155, 157, 158
Lucretius, 15, 24
Lunar Orbiter, 31
Lyell, Charles, 21
Malin Space Science Systems (MSSS), 98, 105, 106, 109, 110, 114, 127, 128, 129, 130, 133, 134, 135
Malin, Michael, 67, 68, 83, 95, 98, 99, 102, 103, 104, 105, 106, 107, 109, 110, 112, 122, 123, 124, 125, 129, 154, 158
Marcahuasi, 170, 171
Marcel, Jesse, 26
Marconi, Guglielmo, 28, 29, 42
Marduk, 7
Mariner 9, 3, 41, 44, 47, 65, 86, 88, 89, 97, 99
Mars Global Surveyor (MGS), iii, iv, 3, 4, 85, 86, 108, 109, 110, 111, 112, 113, 114, 115, 116, 117, 118, 119, 120, 121, 122, 123, 124, 125, 126, 127, 129, 130, 131, 133, 134, 136, 137, 138, 139, 140, 148, 150, 152, 158, 163, 184, 185

Mars Observer Camera (MOC), 98, 107, 108, 109, 110, 134, 139, 152
Mars Odyssey, iv, 150
Mars Orbiter Laser Altimeter (MOLA), 86, 140, 149, 152, 163
Mars Pathfinder, 87, 95, 96
Massachusetts Institute of Technology (MIT), 60
Masursky, Harold, 55, 112, 156
Maunder, Edward, 23, 158
McDaniel, Stanley, 67, 69, 75, 79, 80, 95, 101, 102, 103, 104, 106, 110, 111, 112, 115, 122, 123, 127, 128, 132, 151, 159, 174
McKay, Chris, 55, 90, 91
McKay, David, 78, 92
Menkaure, 167
microbes, iv, 1, 3, 78, 91, 135, 177
MJ-12, 26
Molenaar, Gregory, 1, 45, 46, 47, 49, 50, 51, 56, 59, 65, 70, 96, 97, 102, 106, 123, 157
Moon, v, 3, 5, 7, 8, 9, 13, 15, 16, 17, 18, 19, 31, 32, 33, 34, 35, 36, 38, 41, 42, 44, 57, 82, 105, 123, 153, 157, 182, 183, 184, 185
Moore, Harry, 134, 152
Morrison, Philip, 29, 95
Muslim, 14
Nazca, 181
Nefilim, 5, 6, 7
Nerbun, Peter, 111
Nergal, 7, 169, 170, 173
Ness, Peter, 135, 152, 189
Nineveh, 8, 176
Ninhursag, 7, 173
Ninti, 7
Ninurta, 169

Nippur, 7
Nirgal, 109
Noah, 7, 169, 170
O'Leary, Brian, 60
Ockham, Willian of, 14, 157
Olympus Mons, 41, 109, 164
Oparin, A. I., 93
Oresme, Nicole, 14
Orme, Greg, 135, 152, 189
Owen, Toby, 43
Ozma, 29
Palermo, Effrain, 134, 152
Parker, Timothy, 95, 115, 117, 120
Phobos, 98
Pieri, David, 85, 95, 123, 124, 128, 132
Pilcher, Carl, 110, 128
Plassmann, Joseph, 28
Plato, 11, 12, 23, 31, 187
Plutarch, 16
Popper, Karl, 172
Pozos, Randy, 1, 49, 50, 52, 56, 95, 177
Praxiteles, 57
Ptolemy, 10, 16
Pythagoras, 10, 142
Rautenberg, Thomas, 58
rectilinearity, 105, 154, 179, 182
right triangle, 10, 28, 68, 150
Robertson, H. P., 27, 28, 36, 40, 41
Roe, Robert A., 99
Roswell, 26
Ruzo, Daniel, 170, 171, 175
Sagan, Carl, 2, 5, 6, 23, 25, 30, 31, 34, 36, 42, 43, 44, 55, 56, 57, 58, 59, 62, 63, 65, 77, 79, 95, 103, 105, 112, 123, 124, 136, 154, 156, 158, 174
Saunders, William, 148
Schiaparelli, Giovanni, 2, 22, 23, 155
Schoch, Robert, 168, 169, 175
Schultz, Peter, 164, 165, 175
Search for Extraterrestrial Intelligence (SETI), 5, 29, 30, 110, 123, 124, 156, 157, 172, 177
Secchi, Angelo, 21
seeps, 90, 135
Senapathy, Periannan, 186, 187, 189
Sirisena, Ananda, 119, 161
Sitchin, Zecharia, 5, 6, 7, 8, 10, 23, 170, 172, 173, 174, 175
SNC, 92, 94
Society for Planetary SETI Research (SPSR), 110, 112, 122, 128, 129
Soffen, Gerald, 1, 2, 44, 45, 46, 47, 55, 62
Solon, 11
solstice, 49, 159, 160, 166, 167, 173, 174
Space Shuttle, 38, 39, 98
Sphinx, 167, 168, 170, 175
Spinoza, 153
Star Trek, 63, 177
Starfish Pyramid, 48, 74, 75, 126, 127, 128, 161, 166, 179
Steckling, Fred, 33, 34, 42
Stein, Michael, 63, 79, 104, 179
Stonehenge, 159, 167
Strange, James, 65, 69
Sturrock, Peter, 69
Sumerians, 7, 8, 9

Swerdlow, N. M., 9, 23
symmetry, iv, 51, 66, 70, 71, 103, 105, 119, 120, 122, 136, 137, 140, 141, 143, 144, 147, 148, 150, 151, 152, 154, 158, 179
Teotihuacan, 164, 165
terraformed, 81
Tesla, Nikola, 28, 29, 30, 42
Thales, 9, 10, 11
The Analytic Sciences Corporation (TASC), 58, 60, 62
THEMIS, 150, 151, 184
Tholus, 53, 68, 76, 77, 130, 131
Thompson, Richard, 188
Timaeus, 11, 12, 23, 187
Torun, Erol, 65, 66, 67, 68, 76, 99, 101, 150, 159
tubes, 133
UFO, iii, 26, 27, 28, 37, 38
Ukert, 35, 184
uniformitarianism, 21, 187
United Nations, 42
Urey, Harold, 93
Userkaf, 168
Utnapishtim, 7, 169
Valles Marineris, 41, 86, 109
van Flandern, Thomas, 78, 82, 95, 134, 164, 165, 166
van Hove, Patrick, 60

Verne, Jules, 25
Viking, iii, 1, 2, 3, 4, 41, 42, 43, 44, 45, 46, 49, 53, 55, 58, 62, 63, 68, 71, 72, 75, 76, 83, 88, 91, 92, 96, 97, 98, 99, 101, 103, 104, 106, 109, 111, 113, 114, 115, 116, 117, 118, 119, 120, 121, 122, 123, 124, 125, 126, 129, 130, 131, 132, 133, 136, 137, 150, 161, 162, 166, 167, 177, 179, 184, 185
von Daniken, Erich, 6, 10, 23, 51
von Littrow, Johann, 28
Wall Street Journal, 99, 112
Wallace, David, 43, 44, 56, 77, 95, 105, 136, 158
Ward, William, 95, 160, 174
water, 2, 3, 11, 12, 16, 21, 23, 26, 41, 53, 54, 65, 83, 84, 85, 86, 87, 88, 89, 90, 91, 93, 97, 116, 126, 128, 131, 134, 135, 136, 154, 155, 158, 163, 168, 169, 173
Webb, David, 110, 128
Welles, Orson, 25, 27
Wells, H.G., 25, 41
West, J. A., 29, 168
Wilkins, John, 18
Xenophanes, 10
yugas, 187
Yukteswar, Swami Sri, 189

Notes

Printed in Poland
by Amazon Fulfillment
Poland Sp. z o.o., Wrocław